GALWAY REGIONAL TECHNIC...
CEARDCHOLÁISTE RÉI...

This book s...

Basic Welding and Fabrication

W Kenyon

Principal lecturer/welding and fabrication
Derby College of Further Education

Pitman

021233

PITMAN BOOKS LIMITED
128 Long Acre, London WC2E 9AN

Associated Companies
Pitman Publishing New Zealand Ltd, Wellington
Pitman Publishing Pty Ltd, Melbourne

© W Kenyon 1979

First published in Great Britain 1979
Reprinted 1980, 1982

All rights reserved. No part of this publication may be reproduced,
stored in a retrieval system, or transmitted, in any form or by any
means, electronic, mechanical, photocopying, recording and/or
otherwise without the prior written permission of the publishers.
This book may not be lent, resold, hired out or otherwise disposed of
by way of trade in any form of binding or cover other than that in
which it is published, without the prior consent of the publishers.
This book is sold subject to the Standard Conditions of Sale
of Net Books and may not be resold in the UK below the net price.

Printed and bound in Great Britain
at The Pitman Press, Bath

ISBN 0 273 01321 1

671·52

028895

£5.50

Contents

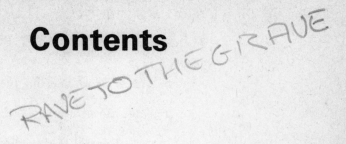

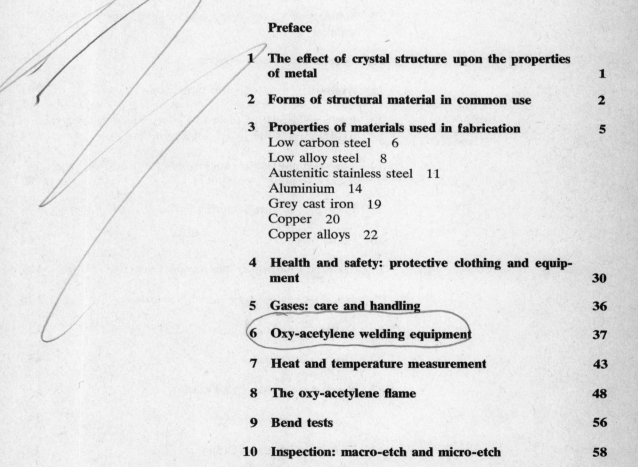

Preface

I have compiled and written this book for those who wish to acquire a basic knowledge of the many common welding and fabrication processes. It aims to provide a good basis for study for students employed in the welding and fabricating industry and following a course of study in further education, such as (in the U.K.) 199F and 200 B.E.C.S., leading to part one of 215, 216 and 217 Craft Studies courses, and some T.E.C. technology units in fabrication and welding.

The first part of this book has been written essentially for information and reference and is not intended as a learning text on material properties and technology. This section may be usefully used in several ways: for example, for reference material to assist in completing the investigations, and for reference with respect to solving practical problems both in the work environment and a training situation. The investigations/demonstrations essentially supplement and reinforce the students information-learning and, with teacher/instructor assistance, they provide practice, which is essential for a true grasp of the principles. The important underlying theme of safety is expressed throughout the book.

The demonstrations are suggested in order that the application of principles can be seen, and follow-up work is provided to encourage the student to think about the practical application of what he has observed. In both Demonstrations and Investigations, typical examples of results are often given, as a guide, to enable the student to draw out in full and complete the various tables, then use them to draw conclusions from the Investigations.

At the end of most sections there is a series of questions designed to test whether the student has understood the content of the section. Some of the questions are multi-choice and require that the correct answer be underlined and others require more detailed answers or sketches.

W. K.

1 The Effect of Crystal Structure upon the Properties of Metal

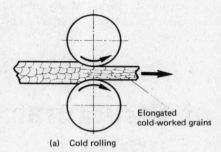

(a) Cold rolling

Elongated cold-worked grains

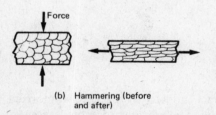

(b) Hammering (before and after)

Fig. 1.1 Effect of cold working

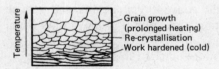

Grain growth (prolonged heating)
Re-crystallisation
Work hardened (cold)

Fig. 1.2 Relationship between grain size and shape, and temperature

All metal is composed of minute particles called crystals or grains. The crystal structure of a pure metal starts forming, around a nucleus, when it begins to "freeze" from a liquid to a solid. The free atoms of a metal become arranged into a regular pattern or space lattice. As heat is extracted from the liquid metal, the crystals grow outwards from each nucleus (dentritic growth) until they form the boundaries of the grains. Provided that the cooling of the hot metal is carried out at a uniform steady rate, the grains should be of a uniform size and shape.

Cold Working and Work Hardening

If a metal is cold worked (e.g. hammered, pressed, rolled, drawn or bent), the crystal structure is altered and the grains become elongated in the direction of working. Fig. 1.1a illustrates the effect of cold rolling and Fig. 1.1b the effect of hammering a metal cold.

This work hardening results in a stiffening of the material, which gradually becomes harder, less ductile and less malleable, but with increased tensile strength. To restore this cold worked structure to fine grains, it is necessary to heat the metal to a temperature at which it will RE-CRYSTALLISE (Re-Cr) and reform into the fine grains.

Care must be taken not to heat the metal to too high a temperature for too long, otherwise the grains continue to grow and become coarse. The metal has then lost its toughness and ductility, and during subsequent forming operations a rumpled "orange peel" effect occurs. Fig. 1.2 shows the relationship between grain size and shape, and temperature.

Hot Working

Hot working takes place above the re-crystallisation temperature of the metal, so that re-crystallisation takes place at the

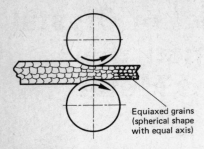

Fig. 1.3 Effect of hot working

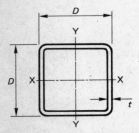

Fig. 2.1 Square hollow section
D = depth of section
t = thickness

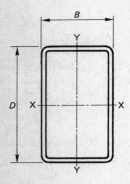

Fig. 2.2 Rectangular hollow section
D = depth of section
B = breadth of section

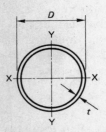

Fig. 2.3 Circular hollow section
D = outside diameter
t = thickness

same time as the deformation, and therefore, work hardening cannot occur. Fig. 1.3 shows the grain flow during hot rolling.

It is important that the hot working is stopped at the correct temperature (just above the Re-Cr temp.) because, if the temperature is too high, grain growth occurs and, if it is too low, work hardening results.

Hot working methods are used to manufacture structural sections. Less power is required to deform the metal since it becomes softer, more plastic and malleable. The hot rolling produces the characteristic fibrous flow of all wrought sections, and it is due to the elongation of the impurities in the original ingot. Forging is also generally a hot working process, because it is essential that the fibres of the metal flow in the direction in which maximum strength is required; this is usually around the contours. Plates, beams, columns, angles, channels and tees are typical hot rolled sections.

2 Forms of Structural Material in Common Use

Beams and Columns

In 1962–63 the Iron and Steel Industry of the U.K. authorised a rationalisation of hot rolled steel sections to achieve greater economy in steel construction. The aims were as follows:
1   To concentrate on universal beams and columns, which are more efficient than the old conventional beams and reduce running costs by permitting the use of plain rather than compound members.
2   To provide lighter and more efficient joists in smaller sizes.
3   To roll channels and angles with square toes in order to facilitate welding.
4   To publicise information on circular hollow sections (designated C.H.S.) and rectangular hollow sections (designated R.H.S.), including square hollow sections.

Note: BS 4 Part 2 designates both imperial and metric equivalents. In Figs. 2.1–2.3, the x and y axis denote the centre of gravity of sections. (The key to symbols is given for sections but not repeated where obvious.)

Rolled Steel Angles (R.S.A.)

Parallel flanges

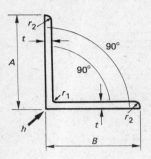

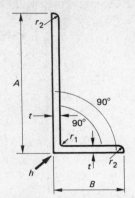

Fig. 2.4 Rolled steel angle
$A \times B$ = size
   r_1 = root radius
   r_2 = toe radius (note square toes)
   h = heel of angle

Fig. 2.5 Rolled steel angle

Bulb Sections

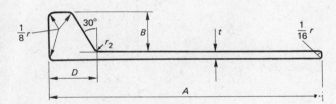

Fig. 2.6 Standard bulb flat (parallel flanges)

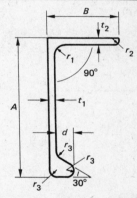

Fig. 2.7 Standard bulb angle
t_1 = web thickness
t_2 = flange thickness
r_3 = bulb radius

Rolled Steel Channels (R.S.C.)

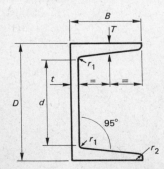

Fig. 2.8 Rolled steel channel
5° taper on flange
D = depth
d = parallel length between fillets
B = flange breadth
T = mean flange thickness
t = web thickness

Rolled Steel Joists (R.S.J.)

5° taper; max. size 203 mm × 101 mm (8 in. × 4 in.)

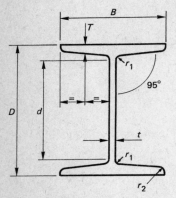

Fig. 2.9 Rolled steel joist

Universal Columns (U.C.)

Parallel flanges

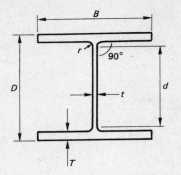

Fig. 2.10 Universal column

Rolled Steel Tees (R.S.T.)

(Note: tees may be cut from universal beams and columns.)

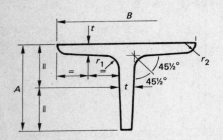

Fig. 2.11 Standard tee
1° taper on stalk

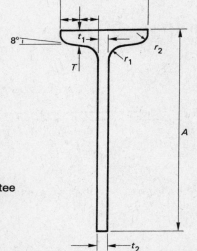

Fig. 2.12 Long stalk tee
8° taper on flange

Universal Beams (U.B.)

2°52′ taper on flange or with a parallel flange.

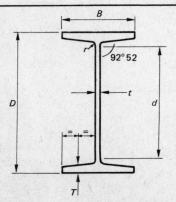

Fig. 2.13 Universal beam

4

Heavy p – g – p

Light p – g – p

Small p – g – p

Fig. 2.14 Super-grip floor plate (steel)

Heavy chequer

Five bar

Fig. 2.15 Tread plate (aluminium)

Plates

Flat

Hot rolled flat plate is rolled in various sizes, both imperial and metric. The plates may be cut to specific lengths and widths to order when the edge is reasonably square, or the plates may still have the mill-rolled rounded edge.

Tread Plate and Floor Plate

The old Admiralty diamond chequer plate has now been replaced by the supergrip floor plate (Durbar) shown in Fig. 2.14 for steels, but it should be noted that heavy chequer diamond, five bar and tread plate (Fig. 2.15) are all rolled in aluminium plate. The basic difference between the two is that steel floor plate is load bearing but not the aluminium tread plate.

(p-g-p = positive grid pattern tread plate.)

3 Properties of Materials used in Fabrication

It is important that all the following materials are compared with each other with respect to their properties, and although specific properties for each material have been given, all are compared with Low Carbon Steel as the base.

The following table gives the symbols of the chemical elements:

Al	Aluminium	Ni	Nickel
C	Carbon	O	Oxygen
Co	Cobalt	P	Phosphorus
Cr	Chromium	Pb	Lead
Cu	Copper	S	Sulphur
H	Hydrogen	Si	Silicon
Fe	Iron	Sn	Tin
Mg	Magnesium	Ta	Tantalum
Mn	Manganese	Ti	Titanium
Mo	Molybdenum	V	Vanadium
N	Nitrogen	W	Tungsten
Nb	Niobium	Zn	Zinc

LOW CARBON STEEL (L.C.S.)

Properties

The following refers to BS 4360 Pt. 2 1969 *Weldable structural steels suitable for bolted and riveted structures* (43A type).
Typical composition C—0.25% max., S—0.05% max., P—0.05% max., Fe—remainder. (May also be supplied to contain 0.2 to 0.5% Cu.)
Approx. melting temperature range 1460°C to 1500°C.
Mass Steel weighs 7.85 kilogrammes per square metre per millimeter of thickness (7.85 g/cm^3 at +20°C).
Hardness 220 Hv. (Vickers Hardness Scale.)
Coefficient of linear expansion 0.000 011 8 per °C.
Tensile strength 430–510 N/mm^2 (43–51 h-bar).
Yield point 220 to 245 N/mm^2 (22–24 h-bar).

Low carbon steel, formerly referred to as mild steel, is the most commonly used material for general fabrication. This is simply because it possesses most of the desirable working properties. These properties are due to the composition of the steel. The iron (ferrite) imparts ductility, malleability, plasticity, magnetism, softness and a certain amount of elasticity, but readily forms oxides, which result in rusting, and heavy scaling when at red heat. The non-metallic element carbon, when added to the iron, produces steel, and is responsible for an increase in hardness and tensile strength. The more carbon added, the higher the tensile strength and hardness, but there is a reduction in the desirable working properties such as ductility and malleability and a greater risk of welds and sharp bends cracking. Normal L.C.S. does not work-harden rapidly but nevertheless cold bending and hammering does cause work hardening, which may be removed by heat treatment. Sulphur and phosphorus are kept to an absolute minimum because they cause a reduction in the desirable properties.

Identification

When steel is filed, it has a light silver grey colour, but when first delivered from the hot rolling mills it is covered by a bluish red mill scale. When sections are allowed to weather in the plate yard, this mill scale turns to the characteristic rust colour.

L.C.S. is magnetic up to 770°C; beyond that, the magnetism is lost through an atomic rearrangement but is regained upon cooling.

Steel may also be identified by its spark colour and the shape of the sparks.

Corrosion

All acids will attack L.C.S. and, wherever there is moisture combined with oxygen, rusting takes place (damp atmosphere). Common salt (sodium chloride) will accelerate corrosion when combined with water, e.g. road salt on cars and pier legs in sea water.

Cutting (Mechanical)

Low carbon steel may be mechanically cut using virtually any cutting machine, provided it is within the capacity of the machine. The capacity is usually stated boldly on the particular machine, complete with a small drawing, where necessary, of the section.

Cutting (Thermal)

Low carbon steel is the material most readily cut by oxy-fuel gas because of its very high iron content. For thick plate and heavy sections, and intricate shapes, it is the most frequently used process. Under correct conditions a very smooth cut edge results. Carbon tends to migrate to the cut edge and upon cooling causes a hard shell to form at the cut edge.

Forming

Low carbon steel has good ductility and plasticity which enables it to be quite easily worked cold into shapes by rolling, flanging, bending or pressing. It has good malleability which enables it to be substantially hammered without too much work hardening, such as levelling plates and riveting. Some *"springback"* due to the *elasticity* is noticed when rolling plates into cylinders. Bending wherever possible should be carried out across the grain for maximum bend strength. If a flange has been bent wrongly and requires straightening and re-flanging, softening may be achieved by annealing and slow cooling.

Pressing and flanging this material hot presents no problems, provided that during heating it is not overheated or "burnt" which causes the material to split and crack during forming. Heating increases malleability, softness and plasticity, and allows maximum stretching and compression without work hardening and cracking, particularly on sharp bends.

Welding

The most commonly used joining process is undoubtedly welding. For the very thick sections, electro-slag, consumable nozzle, submerged arc, electro-gas methods are used. Relatively thin plates are joined using metal arc gas shielded CO_2 and manual metal arc. Thick walled pipes are often given evenly penetrated smooth root runs by using E.B. consumable inserts and the tungsten arc process for fusing the root.

For lap joints on relatively thin plate, resistance welding is used extensively. Oxy-acetylene welding is used on thin sections, especially with car repairs.

Bolting

Black bolts with 1.5 mm clearance are used as service bolts mainly as temporary fasteners. High-strength friction grip bolts and turned barrel bolts, load indicating bolts and washers are used for permanent fastenings.

Riveting

Riveting of low carbon steel is invariably carried out in the shop under favourable conditions, but in most cases has been superseded by welding and bolting, except for repairs on ships and certain low-pressure boilers. Structures which are subject to vibrating loads which also includes a tensile pull, such as transverse stiffeners on bridge structures, are often still riveted.

Because of their favourable properties such as ductility, plasticity and malleability, rivets may be re-tightened if inspection shows that this is warranted. This is the main reason they are still used in specific instances. Bolts tend to become slack or fracture and if a weld cracks it propagates fairly rapidly. Riveted seams tend also to act as crack arrestors. In addition to solid rivets there are many types of hollow rivets used to join sheet metal very quickly and efficiently (see section on rivets).

HIGH STRENGTH AND LOW ALLOY STEEL (L.A.S.)

B.S. 4360 50B and 55 type

Properties

Composition   C—0.20% max., Si—0.50%, Mn—1.5% max., Nb—0.10%, S—0.05% max., P—0.05% max. Plates over 12.5 mm thick supplied in a normalised condition to obtain a combination of strength and notch ductility. Low alloy steel is legibly marked with a light blue paint, i.e. 50B, and if specified, plate will be marked with a diagonal cross or a 12 mm thick line along the length of sections in the same colour paint.
Approx. melting temperature range   Similar to L.C.S.

Hardness Depends upon heat treatment, but if quenched after heating such as welding or thermal cutting, it is several points harder on the Vickers scale, i.e. max. 466 Hv. Normally the average hardness is 200 to 330 Hv.

Tensile strength 500 to 620 N/mm^2 (50 to 62 h-bar).

Yield stress 320 to 350 N/mm^2 (32 to 35 h-bar).

Mass For all practical purposes the same density as L.C.S.

Note: Under the heading of low alloy steels come the *High Yield* and *High Strength* steels as used in the fabricating industry. It is the combination of certain elements in the above steels which are added during manufacture which impart specific properties to the material. A brief outline is now given.

To increase hardenability C–Cr–Ni. (Nb refines grain.)

Toughness (Normal and low temperatures) Ni–Al

Development of corrosion resistance Cu–Cr

High tensile properties C–Mn–Ni–Cr. The carbon is kept to a minimum. Silicon is added as a deoxidiser and also to provide oxidation resistance.

Identification

L.A.S. is similar to L.C.S. in colour, density and magnetism, but with a slightly better corrosion resistance. The L.A.S.s can be satisfactorily worked and shaped using methods similar to those used for L.C.S., with certain exceptions (as noted).

Forming

Cold forming is carried out on sections up to approx. 40 mm thick but, beyond that, hot forming is usually necessary (above 850°C). Greater force is necessary to cold form L.A.S. compared with L.C.S. due to the greater strength of the L.A.S. and greater degree of springback. The ductility of L.A.S. is lower than the L.C.S.

Quenched and tempered alloy steels are limited to a hot working temperature of approx. 650°C, otherwise there is a significant loss of strength unless subsequently heat treated. As L.A.S.s are harder, stronger and tougher than L.C.S.s care must be taken to see that the capacity of the forming machines is adequate. Sharp edges should be removed before forming to prevent cracking.

Cutting

When mechanical cutting, alloy blades should be used with correct clearance and with clean sharp edges. During milling or planing of gas cut edges, it is advisable to obtain a "bite" beneath the hard shell to avoid excessive wear on the tool edge.

Provided the correct conditions are used when oxy-fuel gas cutting, a good smooth profile is obtained. Pre-heating and slow cooling is advisable to prevent hard edges, and indeed in certain cases cracks form at the brittle edge, when cutting thick sections without pre-heat. Certain codes require that between 3 and 6 mm of the gas cut edge is removed to ensure that no hairline cracks remain, and that rolls and forming tools are not scored or damaged by the hard edge.

Welding

The welding processes used are similar to those used on L.C.S. When welding the higher tensile high yield steels, a great problem is what is known as underbead or hard zone cracking. This is cracking that occurs adjacent to the fusion boundary in the heat-affected zone. It very often starts at the root or toe of butt and fillet welds and runs parallel to the fusion boundary, but cracks may be hidden below the plate surface. Cracking may also occur some time after welding and inspection.

The three main causes of cracking are:

a) The presence of hydrogen: from electrode coatings; residual lubricants on welding wire; moisture on plate, etc.

b) The microstructure of the heat-affected zone (H.A.Z.). Rapid cooling from a high temperature produces a coarse grained hard microstructure which is prone to cracking.

c) The high stress level of the joint caused by contraction and restraint.

To prevent H.A.Z. cracking the following is recommended:

a) Keep the hydrogen level as low as possible by using low potential hydrogen electrodes (H) and dry them in an oven to maker's recommendations *immediately* before use. Ensure that all welding wire is clean. Weld-edge preparations should be free from moisture.

b) The cooling rate should be controlled, i.e. use pre-heat up to 200°C with a controlled interpass temperature. Higher pre-heats are used sometimes.

c) Avoid stress build up. The joint fit-up should be controlled. Gaps above 1.6 mm increase the risk of cracking. Welding procedure should be controlled. The more restraint used such as clamping, etc., the greater the stress level. Balanced welding is preferred.

Bolting

Where bolting is specified it is important that high tensile low alloy steel bolts and washers are used. On steel-framed buildings these are usually of the friction grip type. On manhole covers and access ways on steam generating plant they are usually fitted bolts.

Riveting

There are few applications for high tensile rivets. When used they are usually in the form of cold pins of a drive-in fit, mainly used to withstand heavy shear loads such as axle box brackets on rolling stock.

AUSTENITIC STAINLESS STEEL (18/8 S.S.)

Properties

Typical composition   18% chromium, 8% nickel, 0.15% carbon. Remainder: Fe (S. & P. kept below 0.045%.), Ti or Nb.
Effect of elements on properties   The nickel is added to give toughness and may be increased to 11.5% to prevent work hardening (for rivets and fastenings). The chromium is added to give corrosion resistance. In addition titanium or niobium is often added to prevent intergranular corrosion (weld decay) in the following proportions: Ti = 5 × C content, Nb = 10 × C content.
Approx. melting temperature   1420°C to 1395°C.
Hardness   170 Hv (water quench from 1000°C).
Tensile strength   Approx. 620 N/mm².
Yield point Approx. 280 N/mm².
Mass 7.92 g/cm³ at +20°C.
Coefficient of linear expansion   0.000 02 per °C.

Identification

This iron-based alloy has a characteristic silver lustre imparted to it by the amount of chromium and nickel. It is the chromium which combines with oxygen and rapidly forms a very thin oxide which is continuous and stable and impervious to further attack by the atmosphere. The thickness of this oxide film increases with the degree of polish to give a mirror finish. These steels give a dull red spark when touched on a grinding wheel.

Corrosion

The corrosion resistance is excellent in most environments but solutions of nitric, hydrofluoric and sulphuric acid will attack 18/8 stainless steel. These acid solutions are used to remove the oxide scale which result from strongly heating the material in air. The term used for this treatment is *pickling*.

Weld decay (intercrystalline corrosion)   Although 18/8 stainless steel has excellent corrosion resistance, if it is to be situated in a corrosive environment (such as slightly acidic water or corrosive liquid) it requires to be stabilised if welding is used to make the joints. This means that, even with a carbon content below 0.1%, slow cooling of the metal to room temperature after welding will cause the carbon to combine with the chromium to form chromium carbide within the temperature range 500–900 °C. This chromium carbide is then precipitated or thrown out of solution along the grain boundaries. As the carbon has united with the chromium, the iron is now left with little or no chromium near the grain boundaries, and so the iron in these regions is attacked by the acid. Titanium (plate) or niobium (electrode) is added during manufacture because carbon combines easily with these elements rather than with chromium and therefore leaves the iron combined and protected by the chromium. Fig. 3.1 shows diagrammatically the relative area of weld decay in the temperature range.

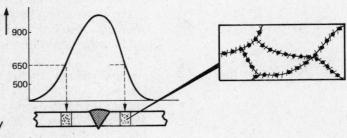

Fig. 3.1 Weld decay

● ● ● ● ● Precipitation of chromium carbides

———— Grain boundary

//////// Chromium depletion (corrosion attack)

Two other types of corrosion to which the 18/8 type of steels are susceptible are: pitting corrosion due to chlorine ions, and stress corrosion due to residual stresses in conjunction with a corrosive medium. Molybdenum is added during steelmaking to reduce these two types of corrosion.

Cutting

All the mechanical methods of cutting stainless steel result in work-hardened edges with sharp burrs, especially if the blade or punch clearances are on the large side. Deburring should be carried out in the interests of safety. The shear strength of 18/8 is higher than L.C.S.; therefore care must be taken not to overload machines which are given capacity ratings for L.C.S. Blade clearances for guillotines are approx. 5% of the material thickness with 1% rake angle.

Excessive clearance results in the material flowing over the bottom blade and so producing excessive burr, a rounded top edge and rapid work hardening in addition to machine overloading. Removing the arris or burr is good practice. Insufficient clearance also causes machine overloading and a poor cut edge.

If the clearance is correct, the bright burnished area of the cut should extend about 40% of the plate thickness. Blades are usually of high speed or alloy steel with four cutting edges which are turned regularly to give a constantly sharp edge.

The requirements for punching are similar for guillotining. The minimum hole diameter is approx. twice the plate thickness and the minimum cross-centres $1\frac{1}{2}$ times the diameter. More distortion is encountered with fully softened material than with cold worked material. The nibbling capacity is about four gauges less than for L.C.S. and the same for cropping. Shearing capacity is approximately half that of L.C.S.

Thermal Cutting

Austenitic 18/8 stainless steel may be thermally cut (using heat) by processes which give varying qualities of cut face depending upon requirements. These methods are briefly summarised below.

1 *Powder Cutting*
Fine iron powder is injected into an oxy-fuel gas flame outside the nozzle. This raises the temperature of combustion and also provides a scouring effect which removes dross and oxides. The result is a fairly rough cut edge and is used for trimming billets and plate slab. Dry air or nitrogen is used to convey the powder (oxygen could cause explosions).

A modification of this method is the Kinox process used for cutting the 18/8 stainless steels. A powder (titanium oxide) is picked up by the oxygen stream and carried to the torch from a powder dispenser. The powder exits through the centre-cutting oxygen orifice, which has a tungsten insert to counter the abrasion. There is no danger of explosion as the powder is inert.

As the powder melts, it provides an abrasive scouring action which removes the refractory chromium oxide which is fluxed with the powder. The velocity of the cutting oxygen ejects the dross from the kerf to provide a reasonably good cut.

2 *Arc Plasma Process*
This process is carried out using an arc which heats a gas to a very high temperature in a water-cooled nozzle. This nozzle constricts the arc, so increasing the velocity and the temperature. The resulting plasma stream blasts and vapourises the stainless steel to form a narrow cut with a slight taper on one edge. The cut is excellent and needs little or no cleaning. See page 133 for further details.

ALUMINIUM

Properties

Colour Aluminium is readily identified by its characteristic polished silver colour. It turns to a light grey due to the oxide formed when exposed to the atmosphere. This film when newly formed is porous and can be coloured by dyes, then sealed (called *anodising*). This oxide is very tenacious and refractory.

Melting temperature Pure aluminium 660°C.

Alloys between 520 and 660°C. (The m.p. of the oxide of aluminium is more than three times this.)

Mass 2.79 g/cm³. Aluminium is extremely light in weight compared with other metals and has very good conductivity. Aluminium is nonmagnetic.

Tensile strength and hardness This varies considerably, from approximately 25 Hv for pure aluminium to over 165 Hv for work-hardened and precipitation-hardened aluminium. Similarly for the tensile strength. See table on page 16.

Coefficient of linear expansion 25.6×10^{-6} per °C.

The ductility, plasticity and malleability of aluminium are very good. It is readily shaped and formed, either hot or cold. During cold working, the material work hardens considerably and annealing needs to be carried out to soften it before further working, or cracking may result. Sections in Al need to be deeper for equal deflection than equivalent steel ones. To anneal aluminium, it is heated to between 350–400°C (a matchstick when drawn on the metal becomes charred to a brown colour), then quenched or allowed to cool in still air.

Toughness

Aluminium has the property of remaining quite tough at very low temperatures. This is one of the reasons it is used for transporting liquid gases at sub-zero temperatures.

Thermal Conductivity and Expansion

The thermal conductivity is about five times that of low carbon steel. It has a coefficient of linear expansion less than 18/8 stainless but about twice that of low carbon steel. This must be taken into account when joining these two materials.

The electrical conductivity is about 60% that of copper but weight for weight it is better than copper. Hence its use nowadays for welding and electrical supply cables.

Corrosion Resistance

The corrosion resistance of aluminium is excellent in certain cases due to the thin film of self-sealing protective oxide. An exception is when aluminium is combined with metals such as steel and copper in the presence of moisture in the form of acid or salt water. This sets up an electrolytic action which results in corrosion of the aluminium rather than of the steel.

Anodising consists of electrolytically depositing a relatively thick coating of a coloured oxide on the aluminium, whilst in an acid bath of sulphuric or chromic acid; it is often used for protecting aluminium.

Barium chlorate or fibrous packing of some neutral substance is often used to insulate two dissimilar metals.

Apart from the commercially pure aluminium, wrought aluminium may be conveniently grouped under the headings:

Non-heat treatable (designated N)

Heat treatable (designated H)

a) The non-heat treatable class is the $1\frac{1}{4}$% manganese type (N3) which may be strengthened by cold working; supplied in 4 tempers of H2, H4, H6, H8; previously known as half hard, three-quarter hard, etc. They may be softened by heating within the range 320°C–420°C (350°C), followed by slow cooling (O condition).

b) The heat-treatable alloys are capable of being age hardened and strengthened with little loss of ductility, by a process known as age hardening by solution treatment. The main alloying elements are small quantities of copper, magnesium, manganese and silicon.

The age hardening heat treatment consists of:

a) Heating the alloy to a prescribed temperature. Depending upon the alloying elements, it is usually between 450°C–540°C so that the elements dissolve into the aluminium.

b) Quenching in water (solution treatment). May be stored in a refrigerator at this stage.

c) Natural aging at room temperature for about a week (T.B. condition), or artificially aging at 160 to 180°C between 2 and 20 hours (T.F.). Heating to much above this temperature causes overaging and a loss in hardness and tensile strength (aging condition 100–200). The strengthening of these alloys is due to the precipitation of sub-microscopic particles of such compounds as Mg_2Si (magnesium silicate) and $CuAl_2$ (copper aluminide), which are trapped within the atomic lattice of the aluminium so strengthening it. Sections may be hot rolled or extruded cold and then heat treated to give the improved properties. Remember, welding always causes softening in the H.A.Z. of the heat-treatable alloys.

Non-heat treated			Tensile Strength	Hardness
Mg 2.25%; Mn 0.3%; Al 97.4%		NG4-0	190 N/mm² Annealed	58 VPN
	Al 99.5%	EIB-O	70.5 N/mm² Annealed	25 VPN
Heat treated				
Cu 4.5%; Mg 0.7%; Si 0.8%; Mn 0.15%; Al 93.8%		HE15TF	465 N/mm² Quenched at 505°C Aged 5 hrs at 185°C	178 VPN

Condition

  M = as manufactured, O = Annealed,
  H = Strain hardened (cold worked)
T.B. = solution treatment and naturally aged
T.F. = solution treated and precipitation treated

Form of material
Forged (F). Rivet (R). Bolt (B). Wire (G).
Sheet or strip (S). Bars or section (E).

Aluminium is particularly resistant to concentrated nitric and acetic acid, but not to alkalis. Any mercury-containing compositions attack Al and its alloys most severely. Flux residues must be removed to avoid corrosion after brazing.

The Al-magnesium alloys with 1% Mn, 0.5% Cr, possess a high resistance to corrosion, especially in sea water and marine atmospheres. The Al-magnesium 5–7% alloys are prone to intercrystalline corrosion especially when highly stressed.

Cutting (Mechanical)

Aluminium may be mechanically cut or punched on any machines used for low carbon steel. The blades and cutting edges must be scrupulously clean to avoid contamination of the cut edge with iron particles. The clearances need to be altered for equivalent thicknesses where necessary but contours may be nibbled out at thicknesses below approximately 22 mm. In addition the soft aluminiums may be machined using high speed routing machines.

Cutting (Thermal)

The thermal cutting of aluminium is usually done using the arc plasma process. In this process an arc heats a gas to a very high temperature in a water-cooled nozzle. This nozzle constricts the arc, so increasing its velocity, and combined with the high temperature, the resulting plasma stream blasts and vapourises the aluminium to form a narrow cut. A slight taper on the cut edge is an unavoidable feature. For thin sections the laser and electron beam are sometimes used. See pages 136 and 139 for further details.

Forming

Aluminium may be spun, rolled, bent or pressed similarly to L.C.S. but may require to be softened in between the forming stages due to work hardening. When setting plate edges in a press brake, prior to rolling into cones or cylinders, care must be exercised in applying pressure. Aluminium is soft and easily marked. Indentation marks and ripples may result which are unsightly and make cleaning difficult.

Welding

The most common joining process for thin and thick sections of aluminium and its alloys is fusion welding. The metal arc gas shielded process using argon as a shielding gas is usually used for materials above 6 mm thick for production work. The tungsten arc gas shielded method is used for less than this thickness and for pipes, but this is not a hard and fast rule and depends upon the particular job. Aluminium flame brazing and friction welding are also processes which have specific applications. A three-phase power supply is required when resistance welding aluminium. Thin sheet may be welded using plasma arc or electron beam.

Some important points to note when fusion welding aluminium are:

a) The high conductivity increases the width of the H.A.Z. and also necessitates pre-heating in certain cases.

b) The high thermal expansion causes more distortion than L.C.S.

c) The welding causes the work-hardened or the heat-treated material to lose some of its tensile strength and hardness in the zone affected by the heat from welding.

d) All the Al–Mg–Si and Al–Mg–Zn alloys are welded using non-matching fillers to avoid solidification cracking. Al–Mg–Si (filler Al 10% Si); Al–Mg–Zn (filler Al 5% Mg).

e) A zirconiated tungsten is used for welding aluminium alloys.

f) The Al magnesium alloys are sensitive to exaggerated grain growth in the H.A.Z.

g) Porosity is sometimes a problem because of gas inclusion. All joints should be cleaned of grease, dirt and moisture by solvents (acetone) and oxides removed with stainless steel wire or an inert scouring material.

h) There is no colour change of Al upon heating.

Brazing

Furnace or torch brazing in the range 550–620°C using a 5 to 12% silicon rod. A flux based on mixtures of alkali metal chlorides and fluorides is a cheap rapid method of joining aluminium, especially for lap and fillet joints.

Bolting and Riveting

(BS 1473) 12 to 25 mm dia. Rivets must be of the same specification as the parent material to avoid galvanic corrosion and may be solid, tubular or "blind" rivets. They are usually driven cold and, if of the age hardening type, may be kept refrigerated just prior to driving. Pneumatic or squeeze riveting is usually employed for forming the head.

Bolts of the high-strength heat-treatable type are used with washers. Bolts of the 18/10 austenitic stainless type are also used.

Adhesive Bonding

Aluminium is an ideal material for bonding by adhesives. The adhesives may be of the thermosetting type, which utilises a chemical reaction assisted by heat and pressure to harden and bond (cure) the epoxy resin to form the joint. They have a high bond strength but tend to be brittle and cannot be remelted.

The other common type is thermoplastic. They may be hardened by cooling and softened by heating, and provided the decomposition temperature of the resin is not exceeded this process can be repeated. Heat is *not* required to form the bond but pressure is needed. Impact types of adhesive are thermoplastic.

The advantages of adhesive bonding may be summarised as follows:
1) Overaging does not occur and there is no distortion due to heat input.
2) It produces smooth joints with no protrusions.
The adhesive bond has low peel and cleavage strength. Joints should be designed on a lap/shear principle with a thin adhesive layer of maximum joint area which is chemically cleaned.

Forms of Supply

(See also page 5).
a) Hot rolled plate over 3 mm thick (may be given a cold rolled finish for superior finish and accuracy).
b) Extruded and rolled sections.
c) Patterned sheet (tread plate), heavy chequer five bar and P–G–P.

GREY CAST IRON

Properties

BS 1452 specifies seven grades, numbered according to minimum tensile strength of the material.

Composition   Carbon 2 to 4%, silicon, iron.

Colour   Silver grey.

Melting temperature   Depends upon the carbon content, approx. 1150 to 1200°C.

Hardness   Varies with cooling rate. In annealed condition between 142 and 325 Hv.

Magnetic   Due to the ferrite (iron) content.

Mass   The density is determined by the graphite content 7050 kg/m³ to 7300 kg/m³.

Coefficient of thermal expansion   Varies between $10–12 \times 10^{-6}$/°C, between 20–200°C.

Conductivity   A medium conductor of heat and electricity.

Tensile strength   Fairly low compared with steel. The seven grades vary between 150 and 400 N/mm² and, due to elongated flakes of graphite (carbon), it has poor bending strength and tends to fracture in a brittle manner quite suddenly. A factor of safety of 4 is frequently used to determine maximum design stress.

Grey cast iron has very poor ductility, plasticity, toughness, malleability but its compressive strength is excellent. The compressive strength is four times tensile strength; shear and torsional strengths are approx. 1.1 tensile strength. Its ability to absorb and dampen vibrations makes it an excellent material for heavy machine bedplates. The more graphite present, the lower the melting point. The free graphite not only dampens vibrations but it also acts as a lubricant on surface tables for marking-out. The excellent castability of this material has also provided its name. Grey cast irons are extremely notch sensitive. Heating between 700–800°C causes a volume increase due to the formation of iron oxide resulting in cracking ("grain growth").

Corrosion Resistance

It has a very good corrosion resistance to the atmosphere, sea water and weak acids.

Cutting (Mechanical)

Grey cast iron may be readily drilled, sawed, or machined but due to its brittle nature cannot be sheared or punched. Chipping or filing are popular methods of preparation along with grinding and routing.

Cutting (Thermal)

Grey cast iron is cut successfully using oxy-fuel gas equipment but a slightly different technique is used from the conventional procedure. The fluid oxides of iron are used to wash away the graphite by a weaving action of the blowpipe, producing a rather wide kerf but an acceptable cut. Arc–air gouging is not so successful but coated electrodes can gouge successfully but with much fume and more dross.

Forming

May be cast in moulds to form quite intricate shapes and, for large numbers off, it is often more economical than fabrication.

Joining

May be readily fusion-welded by m.m.a. using electrodes, which produce either non-machinable welds because of the rapid melting and quenching, or machinable welds by using nickel base electrodes. Bronze welding is a common joining method and so is fusion welding using oxy-acetylene and high silicon filler rods to give a colour match. It is advisable to pre-heat thick sections or irregular sections and slow cool.

Bolting

Sections may be bolted provided no bending stresses are induced, using ordinary grade steel bolts. Riveting is not normally used.

Note If magnesium is added at the manufacturing stage, the graphite, instead of forming flakes on cooling, forms spheres or nodules. This is called S.G. iron (spheroidal graphite) and is used for pipelines because of its increase in ductility and other desirable properties.

COPPER

Colour A reddish pink when polished but forms a brown surface oxide when heated.
Melting temperature 1083°C
Hardness Varies with the degree of cold work.
In the annealed condition 47–52 Hv.
After a fair amount of cold work 85–105 Hv.
Non-magnetic
Mass 8.94 g/cm³ at 20°C.

Coefficient of linear expansion   Copper: 17.7×10^{-6} per °C.
Brasses: $70/30 = 19.9 \times 10^{-6}$ per °C; $60/40 = 20.8 \times 10^{-6}$ per °C.
Conductivity The oxygen-free high-conductivity copper (O.F.H.C.) has extremely high electrical and thermal conductivity. When elements are added to pure copper, the electrical conductivity is lowered.
Tensile strength
Annealed condition 220–250 N/mm²
After cold working 310–400 N/mm²

Copper is very soft in the annealed condition but rapidly becomes harder during cold working, e.g. hammering, rolling or pressing. The increase in hardness is accompanied by an increase in tensile strength but a loss in ductility.

The ductility, plasticity and malleability of pure copper is very good, but requires annealing to remove cold working effects. In the annealed condition it is excellent for deep drawing, extrusion and drawing into wire. Many copper alloys have poor ductility between 400–700°C ("ductility trough").

To anneal copper, heat to a dull red approximately 500°C and allow to cool in air or quench in water to remove oxide scale.

REGIONAL TECHNICAL COLLEGE GALWAY
LIBRARY
028895
20 DEC 1985
671.52

Corrosion Resistance

Copper has a high corrosion resistance to most acids but is attacked by oxidising acids such as nitric and hydrochloric quite vigorously. No copper alloys are compatible with ammonia.

Electrochemical corrosion happens when two dissimilar metals are together in an acidic solution, causing one metal to corrode at the expense of the other. One metal acts as a cathode and the other metal which corrodes is the anode. When copper or its alloys are in contact with other metals in corrosive environments, electrochemical corrosion is a significant factor. A combination of aluminium and copper should be avoided if possible, as the copper is highly cathodic to aluminium and causes the aluminium to corrode rapidly.

Cutting (Mechanical)

Copper in its soft annealed condition tends to clog saw-teeth and takes the sharp cutting edge off tools. This results in large burrs and distortions at the cut edge when shearing and punching. The same difficulty is not experienced when shearing the 70/30 brasses or when machining the 60/40 brasses.

Cutting (Thermal)

Copper may be thermally cut by the arc plasma process using argon/hydrogen or nitrogen mixtures or by laser beam.

Forming

Copper and the 70/30, 65/35 brasses may be spun, rolled, pressed and bent but may require softening by heat treatment in between stages to remove any work hardening effects at approximately 500–600°C.

Joining

Copper and its alloys are readily fusion welded using m.m.a. and copper–tin electrodes, or T.I.G. welded with electrode negative polarity d.c. with argon/helium shielding gas and fillers shown in BS 2901 Pt. 3 and Pt. 2. Other welding methods are M.I.G., friction, cold pressure, oxy-acetylene, plasma arc.

Bolting and Riveting

Bolts and stays are used of the type which will not cause an electrochemical action and are often plated steel, or alloys of copper, such as Monel. Rivets for thin sheet copper are usually of soft commercially pure copper.

COPPER ALLOYS

1 Tough Pitch Copper

Oxygen-bearing copper known as "tough pitch" contains small amounts of oxygen in the form of cuprous oxide. It is used widely for electrical work. It is not readily weldable because of its liability to crack when hot, called "hot shortness". During welding large amounts of porosity due to the formation of steam from oxygen and hydrogen are present.

2 Deoxidised Copper

Phosphorous is added as a deoxidant during manufacture which improves the mechanical properties and reduces hot shortness, but reduces the electrical conductivity. This type of copper is weldable.

3 Arsenical Copper

Adding 0.3 to 0.5 arsenic increases the tensile strength and toughness, especially at temperatures between 200–300°C. The resistance to oxidation and scaling is reduced, hence the reason for use on fireboxes.

The deoxidised type is used when welding, with the tin bronze type electrodes with a 300–400°C pre-heat above 5 mm thick and T.I.G. welding using a copper filler as per BS 2901 and an argon-helium or nitrogen shield. The flow rate for nitrogen is 50% greater than argon. Pre-heat aids fusion on materials above 1.6 mm thick.

Brasses

There are several types of brass, all of which have zinc added in some proportion to the copper. Brass is a golden colour and similar to copper. It varies in tensile strength and hardness depending upon the amount of cold work it is subjected to. It is readily cast, forged or stamped, and extruded or hot rolled (60/40). It may be drawn, pressed, spun or machined, and cut using standard fabrication machines. Lead is sometimes added to improve machining, but this causes some difficulties in welding due to the lead causing porosity. Arsenic is added to reduce "dezincification" and small quantities of aluminium improve corrosion resistance.

The two main groups of brass are:

1 *Cartridge Brass*

70% Cu, 30%Zn	650–600°C Annealed	Cold worked
	U.T.S. 350 N/mm^2	590 N/mm^2
	Hardness 50 Hv	90 Hv

This brass has very high ductility and strength and is used when deep drawing operations need to be carried out. This should be followed by stress relieving which involves heating to approximately 250°C to prevent subsequent cracking in service, often called "season cracking". It has a melting point of 920°C and a good corrosion resistance to sea water and atmospheric corrosion. It is readily soft and silver soldered, brazed and welded using an oxidising flame for welding to prevent zinc loss. Arc welding is as for copper, but T.I.G. welding is carried out using an argon shield and a silicon bronze type rod.

Main uses Deep drawn components such as cartridge cases, spun components for decorative work and some instruments, radiator tanks and cooking utensils.

2 *Muntz Metal*

60% Cu, 40% Zn	Annealed	Cold worked
	U.T.S. 480 N/mm^2	500 N/mm^2
	Hardness 89 Hv	143 Hv

This material works well in the range 650°C to 750°C and is not intended to be coldworked for any significant amount. Welding and brazing are similar to 70/30 brass.

Main uses Extruded sections such as bar or tube. Cast products such as regulator parts including hot stamped components. Hot rolled sheet.

Materials Investigation 1 Comparative Tests on the Working Properties of Low Carbon Steel and High Yield Steel

THEORY   The majority of fabrication work is manufactured from low carbon steel but fabrications which are subject to higher stresses are very often fabricated from high yield steel, e.g. portal frames in buildings or pressure vessels.

The following properties are present to some degree in both materials.

Toughness   The resistance of the material to impact or shock loads. (Sharp notches drastically lower the toughness.)

Tensile strength   The ability to withstand being pulled apart.

Malleability   The property which enables a material to be *compressed* in all directions by either hammering or rolling, without cracking.

Ductility   The property in a material which enables it to be *bent* so that the outside of the bend (which is in tension) does not show signs of cracking. Materials must have ductility to be *drawn* into wire.

Elasticity   The property which enables a material to be stretched, and when the load causing the stretching is removed, to return to its original length. (This property is the cause of "springback" in materials.)

Plasticity   The property which enables a material to be shaped or deformed and retain that shape after the deforming force has been removed (increases upon heating).

Hardness   The ability of a metal to withstand abrasion, scratching, wear, penetration, cutting and machining.

MATERIALS

One 12 mm dia. low carbon steel rivet, one 12 mm square of low carbon steel plate, and one 12 mm square low alloy steel plate.

Two lengths, each 150 mm × 25 mm × 3 mm, both low alloy and low carbon steel plate.

One 6 mm dia. rivet, approx. 50 mm long.

PROCEDURE

A) Place one piece of each material in the folder with the sharp shear edge down. Fold to the 90° mark and test for springback with a square and record. Note any cracking on the sharp outside edge. Repeat with the other specimens, but with edge upwards.

B) Heat the 12 mm squares of plate in the forge with the 12 mm dia. rivet. Using a thick plate with a hole in it and a heavy hammer, flatten the rivet head until it is $2\frac{1}{2} \times$ dia. of shank. Then flatten each piece of plate until it is half its original thickness. Look for cracking on the outer edges.

C) Now bend the 6 mm rivet shank COLD until the shank touches the head and look for cracking on the OUTSIDE of the shank.

QUESTIONS

Now draw and complete a table as shown below. Use the following code to show the properties that you discovered:

M = malleability; D = ductility; E = elasticity; P = plasticity.

	L.A.S.	L.C.S.
A) With the sharp edge down, did cracking result? YES/NO With the sharp edge up, did cracking occur? YES/NO		
	12 mm rivet	6 mm rivet
B) and C) Did the outside of the flattened rivet crack? YES/NO Did the outside of the bent rivet crack? YES/NO		

1) Which property causes springback?
2) Which property allows the metal to remain in the bent position after the force is removed?
3) Which property allows the metal to be bent without cracking?
4) Which two properties do rivet heads possess when hot?

Materials Demonstration

AIM To note the effect of heat treatment on the mechanical properties of plain carbon steel knitting needles.

THEORY Steel is an alloy of the two elements carbon and iron. With a carbon content of 0.6% in its QUENCHED hardest condition, it would be too BRITTLE for any practical use. The operation of TEMPERING within the temperature range 150–300°C is performed to relieve stress without seriously affecting the hardness. Heating above this tempering temperature results in some loss of hardness, *but* an increase in toughness. The steel is at its toughest when allowed to cool slowly in still air from approximately 840°C (light cherry red).

APPARATUS Three 0.6% carbon steel knitting needles, glass tank of water, bunsen burner, pliers, cloth, emery paper.

PROCEDURE

STAGE 1

Slightly bend the first needle to show springback.

Q. Which mechanical property allows the needle to return to its original shape after the bending force is removed?

Q. Give examples of where this property is usefully employed (e.g. rail tyres, springs for rail and auto use, wire rope, riveting snaps).

STAGE 2 *Quenching*

Heat the second needle in the central portion to light cherry red (approximately 840°C), holding one end with pliers, and quench in the tank so the class may see the scale flaking off. Remove after testing for cooling under water, and snap in the cloth.

Q. Which mechanical property allows the needle to fracture due to the small bending force?

Q. Name another material which possesses this property.

Q. Is this a desirable property of engineering materials?

STAGE 3 *Quench and Temper*

Heat and quench the third needle as Stage 2, and then polish with the emery paper. Re-heat the central portion and temper at approximately 300°C (Blue).

Demonstrate springback.

STAGE 4 *Normalise*

Heat the first needle to approximately 840°C (light cherry red) and allow to cool in still air. When cool, bend slowly.

Q. What mechanical property allows the needle to remain in the bent position after removal of the bending force?

Materials Investigation 2
The Effect of Heat Treatment on Specimens of Plain Carbon Steel

APPARATUS

Bench vice, bending bar, and two 5 mm dia. steel specimens

Specimen 1: Low carbon steel, 0.1% carbon (low carbon steel)

Specimen 2: High carbon steel, 1.0% carbon (silver steel)

PROCEDURE

TEST 1   File the end of each specimen.

Complete a table like the one below.

Low carbon steel Result:

High carbon steel Result:

TEST 2   Grip 25 mm in the vice jaws, place the bending bar over the specimen, and bend to 90°.

Low carbon steel   Result:
High carbon steel   Result:

TEST 3   Heat both specimens to a cherry red (approx. 800°C) and quench in water. File the end of each specimen.

Low carbon steel   Result:
High carbon steel   Result:

TEST 4   Grip specimens in vice jaws, place bending bar over the specimen and attempt to bend each specimen to 90°.

Low carbon steel   Result:
High carbon steel   Result:

TEST 5   Polish the high carbon steel specimen with emery paper. Heat the end of the specimen and temper the steel at 280°C (Dark purple).

File the tempered end   Result:
   of the specimen
Attempt to bend to 90°   Result:

QUESTIONS
Some of the following statements are correct, some incorrect.
1)   Low carbon steel can be hardened by simply heating to a cherry red and quenching in water.
2)   Heating a high carbon steel to cherry red and quenching in water will make the steel very hard.
3)   A high carbon steel is harder to file than a low carbon steel.
4)   A soft steel bends more easily than a hardened steel.
5)   A hardened steel is brittle.
6)   A hardened steel can be tempered by heating it up to 600°C.
7)   A brittle steel bends easily.
8)   Tempering makes a steel tougher and less brittle.

The Spark Test

This test is very useful in the workshop to identify a steel and to determine approximately the carbon content. A sample should be held lightly on a grinding wheel in order to observe the shape, colour and length of the streaks and sparks produced.

There is a gradual change in the form of the streak and spark through the series of carbon steels. Wrought iron (very low carbon content) is characterised by the bright yellow streaks, broadening out before disappearing in a curved tail. With increasing carbon content all the following changes occur: (1) the colour of the streak is less bright, (2) the leaf becomes smaller, (3) the spark becomes larger, (4) sparks occur nearer the grinding wheel. High carbon steel is characterised by a profusion of sparks commencing very close to the grinding wheel.

Low carbon steel and medium carbon steel are not very easily identified singly, but when tested together it is comparatively easy to see which has the lower carbon content. A sample of good quality low carbon steel should be kept handy for making direct comparisons with other similar steels.

Malleable iron at first produces sparks similar to wrought iron. When the decarburised skin is penetrated, the sparks are like those of cast iron.

Case hardened steel is high carbon steel on the surface. The carbon content gradually diminishes as the case is ground away.

Stainless steel containing an appreciable quantity of carbon will show the carbon spark at the end of the characteristic bright yellow streak.

Fig 3.2 summarises the tests in pictorial form.

Materials Investigation 3
Simple Recognition Tests and Identification of Materials

THEORY   There are various simple methods for recognising and identifying materials. The most common methods are as follows:

1   Colour
2   Mass (weight), relative density
3   Magnetic properties
4   Hardenability
5   Spark test
6   Melting temperature
7   Action of acids on the material
8   Action of alkalis (sodium hydroxide).

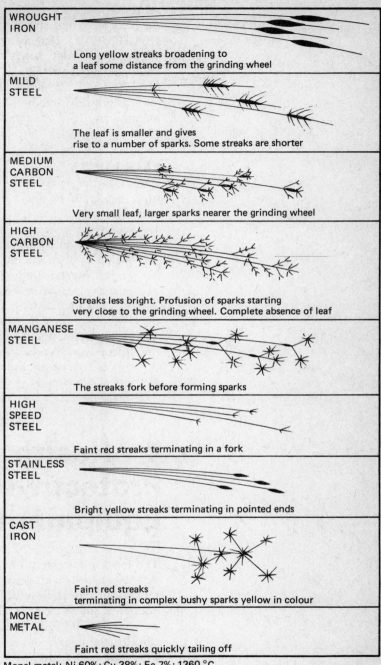

WROUGHT IRON
Long yellow streaks broadening to a leaf some distance from the grinding wheel

MILD STEEL
The leaf is smaller and gives rise to a number of sparks. Some streaks are shorter

MEDIUM CARBON STEEL
Very small leaf, larger sparks nearer the grinding wheel

HIGH CARBON STEEL
Streaks less bright. Profusion of sparks starting very close to the grinding wheel. Complete absence of leaf

MANGANESE STEEL
The streaks fork before forming sparks

HIGH SPEED STEEL
Faint red streaks terminating in a fork

STAINLESS STEEL
Bright yellow streaks terminating in pointed ends

CAST IRON
Faint red streaks terminating in complex bushy sparks yellow in colour

MONEL METAL
Faint red streaks quickly tailing off

Monel metal: Ni 60%; Cu 38%; Fe 7%; 1360 °C

Fig. 3.2 The spark test
(*Courtesy British Steel*)

AIM To identify various materials using the above methods (to include coated sheet).

EQUIPMENT Magnet, dilute hydrochloric acid, weighing apparatus, heating torch and quench tank, grindstone, file, hammer and centre punch.

MATERIALS 25 mm square, one of each—copper, brass, low carbon steel, medium carbon 0.6% C steel, low alloy steel, aluminium, 18/8 austenitic stainless steel, grey cast iron. (Broken Hounsfield tensile test pieces.)

PROCEDURE

Carry out the tests, in the order shown, on each material in turn. While doing so, complete a table of the sort shown opposite.

1) Judge the colour as received and then the colour after cleaning with rough emery cloth.
2) Weigh each specimen.
3) Try for magnetism with the magnet.
4) Test hardness with file then heat and quench and test again to see if the hardness is greater or less.
5) Test each specimen on the side of grindstone held with pliers (CARE) and wearing clear glasses.
6) Record time it takes for each specimen to just melt with an oxy-acetylene torch held in the centre.
7) Put all specimens in dilute acid and then check at weekly intervals for results. The same with sodium hydroxide.

4 Health and Safety: Protective Clothing and Equipment

The legal requirement of the new Health and Safety at Work Act 1974 has created for both employers and employees a more acute awareness of the need to take care in avoiding accidents, injury and disease. The Act states: "It shall be the duty of EVERY employee while at work to take reasonable care for the health and safety of himself and of other persons who may be affected by his acts or omissions at work" (e.g. screening of the arc, wearing goggles, grinding and chipping away from others, replacing guards, keeping gangways clear, marking hot metal, etc.).

It is the duty of the employer to ensure that adequate protective equipment is available, and that adequate guarding of machines is maintained so that they are SAFE when PROPERLY USED (mechanical and photo-electric).

Metal in The Annealed Condition

Action	Example Copper	Brass	L.C.S.	M.C.S.	Cast Iron	Aluminium	18/8 S.S.	L.A.S.
1 Colour as received	Reddish brown							
after clean	Red/pink							
Mass at 20°C g/cm³ Relative density	8 94							
2 Mass at 20°C kg/m³	8940							
3 Magnetised No/Yes	No							
4 Hardness (1 soft to 7 hard)	2							
Hardenability (L/S/M) (L less, S same, M more)	S							
5 Spark test	None							
Melting temp. (°C)	1083	850–950	1450	1425	1175–1250	660	1425	1450–1470
6 Melting time (sec) (oxy-acet. torch)	25 secs							
7 Acid (nitric) 1 day	vigorous							
1 week	vigorous							
8 Alkali 1 day (sodium hydrox.)	no effect							

The employee (you) is required by law to wear certain protective devices in designated areas, for example eye protection. In addition to protective equipment such as helmets, goggles, spectacles, visors, ear muffs (noise), gloves, fireproof aprons and spats, toetector boots, etc. there are other factors to be considered to ensure your health and safety.

Factors to be aware of to avoid injury

1 Never line up holes with your fingers.

2 When using rotating machines (drills), beware of hanging hair, ties, clothing, scarves or belts getting caught.

3 Lift weights correctly with a straight back using the strength of leg and beware of sharp edges (gloves). Rings should not be worn.

4 Know where the first aid room, the nearest phone, the fire alarm and fire extinguishers are.

5 Lift gas cylinders with rope slings and do not use as rollers.

6 Be aware of electrical hazards such as bare wires, poor earth return connections, wet floors (use duck boards to stand on).

7 Do not use oxygen as a substitute for compressed air and never use as a "sweetener" in compartments where the air is stale.

8 Do not use compressed air to blow down clothes as the pressure can cause serious injury to eyes, ears and internal organs.

9 Stack plate, sheets or components tidily and not too high.

10 NEVER remove guards whilst a machine is running (it is illegal) and know where the STOP button is. Do not start a machine without guards in position.

11 Keep your work place clean and tidy and don't use defective tools.

12 Do not fool around in the workshop and always walk not run.

13 Know the warning signs and safety colours and watch out for them.

14 Know your crane signals (see Fig. 4.1).

15 Use ventilation equipment to avoid dangerous concentrations of oxides of nitrogen, ozone, fumes from metals such as lead, zinc and cadmium. The following solvents also give off dangerous vapours: benzene, carbon tetrachloride, trichloroethylene and perchloroethylene (poisonous phosgene gas). Never allow the following to come in contact with the skin: lead paint, corrosive acids, paraffin or oils.

16 Learn the different types of fire extinguisher.

CIRCLE	◯	An order
TRIANGLE	△	Caution
RECTANGLE	▭	Safety information

Red	Obstruction
Orange-yellow	(Tiger stripes Hazard with black diagonals)
Green	Safe route ⇐

Fig. 4.1 Crane signals
(*Courtesy East Midlands Division, National Coal Board*)

Signaller should stand in a secure position where he **CAN SEE the LOAD** and **CAN BE SEEN CLEARLY** by the driver

Face the driver if possible. Each signal should be distinct and clear.

Fire Fighting

The following Table shows the three main classes of fire and the approved fire extinguishing media.

Class of Fire Risk	Type of Extinguishing Agent	Notes
Class A Wood, cloth, paper and similar solid combustible material	Water from fire buckets, hose reels, extinguishers of the gas pressure and soda/acid types	This employs the cooling-down method which reduces the temperature involved to below that of combustion
Class B Flammable liquids—petrol, oils, greases, fats	Asbestos blankets	Suitable for small fires, e.g. frying pans
	Dry powder	Suitable for fires in spilled liquids when quick control is required
	Carbon dioxide (CO_2)	Quick acting, clean, does no damage and is non-conductive
	Foam	Smothers the fire and seals against re-ignition
Class C Electrical equipment	Dry powder Carbon dioxide (CO_2) Vaporising liquids	All are non-conductive. Water *Must Never be Used*. Carbon dioxide causes no damage to delicate equipment

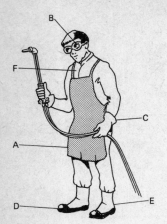

Fig. 4.2

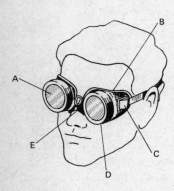

Fig. 4.3

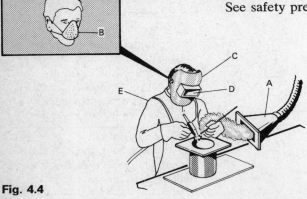

Fig. 4.4

Protective Clothing and Equipment

Figs. 4.2, 4.3, 4.4 illustrate examples of the use of protective clothing and equipment.

FIGURE 4.2

Name	Main Use
A. Flame-resistant apron	Prevents burning of clothes.
B. Gas welding/cutting goggles	Protects eyes from sparks.
C. Gauntlets	Prevents skin burn.
D. Safety boots (steel toecap)	Prevents crushing of toes.
E. Spats	No molten metal down boots.
F. Boilersuit	Protects neck and chest.

FIGURE 4.3

Name	Use
A. (i) Clear glass	Protects tinted lens.
A. (ii) Tinted lens	Limits glare.
B. Goggle body	Stops sparks.
C. Air vent	Prevents misting up.
D. Lens holder	To change broken lens.
E. Strap adjustor	Adjust for size

FIGURE 4.4

Name	Use
A. Extractor fan	Takes away fumes.
B. Filter mask	Dust and fumes.
C. Head shield	Prevents skin burn.
D. (i) Renewable clear glass	Takes spatter, etc.
D. (ii) Renewable tinted	Prevents arc eye.
E. Leather cape with sleeves	For overhead work.

The filter mask (Fig. 4.4B) is no protection from dangerous gases such as phosgene (which is formed from degreasing agents such as trichloroethylene) or nitrous fumes (caused when large areas of plate are heated) or any other poisonous gases. See safety precautions for T.A.G.S. and M.A.G.S. (p. 95).

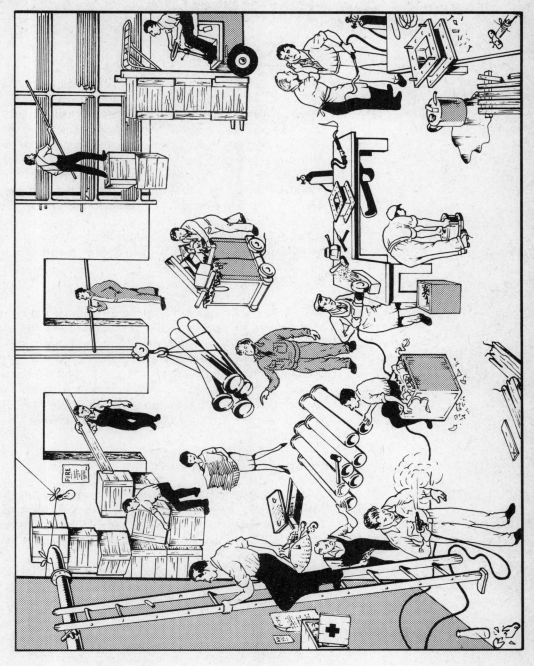

Fig. 4.5 List as many instances of unsafe practice as you can detect (*Courtesy "What's Wrong" Safety Posters*)

5 Gases: Care and Handling

Storage

Cylinders should be stored in a cool dry place away from excessive heat or corrosion. Acetylene cylinders should always be stored upright in a place vented to atmosphere with flame-proof switches and lights. Empty cylinders should be kept separate and clearly marked to distinguish from full ones. Acetylene cylinders should not be mixed with oxygen cylinders. No smoking or naked lights are allowed in the store, which should be easily accessible, with fire extinguishers available (carbon dioxide type). Round bottomed oxygen cylinders may be stacked four high with wedges in either side.

Safe Handling

Cylinders should not be dropped from a height or used as rollers, work supports or jacks. A rope sling should be used for lifting compressed gas cylinders, or special apparatus. Cylinders should be used upright and safely secured away from direct arcing on the cylinder or a naked flame. Do not allow grease or oil to come in contact with cylinders, especially compressed oxygen, as this may cause an explosion. Hydrogen is released from the breakdown of the oil and causes the explosion.

Distribution

Acetylene and Apachi gas should not come into direct contact with copper or alloys with above 70% copper, as an explosive compound (copper acetylide) is formed. For this reason, the acetylene piped from the manifold is conveyed in iron or steel pipes.

Some gases may be delivered as liquid bulk in large tankers and piped as a liquid into the firm's own storage tanks. Examples are oxygen, argon, propane and methane.

Explosions

Some causes of explosions when using gases:

1 Leaking cylinders, torches, gauges and connections.
2 Inadequate ventilation in confined spaces.
3 Transferring gas from one cylinder to another.
4 Dropping and mishandling cylinders and equipment.

5   Leaving lighted torches unattended.

6   Using compressed oxygen as a means of ventilation or compressed air.

7   Allowing hot metal and sparks to fly on to hoses and connections.

8   Testing for leaks with a naked flame.

Flashback may be prevented by fitting pre-flash flashback arrestors which prevent backflow of gases by means of a non-return sintered stainless steel flame trap. Hose check valves fitted at the blowpipe end have a spring-loaded valve which seats instantly the gas flow is reversed.

Identification Colours of Gas Cylinders

The principle of using a colour scheme for identifying cylinders holding gases in common use is that

yellow should represent toxic or poisonous gases and

red or **maroon** inflammable gases.

Some gas cylinders have a distinguishing colour band painted around the neck or down the length. Manufacturers often paint an aluminium panel on the cylinder body to show up special markings, and also attach identification labels.

6   Oxy-Acetylene Welding Equipment

High-pressure Oxy-Acetylene Welding Systems

Consist of the following equipment:

1   A supply of dissolved acetylene (D.A.) stored in steel cylinders which contain a porous substance (Charcoal) and a solvent (acetone) for the gas. The cylinders are charged to a pressure of approximately 15.5 bars. There are various sizes, the usual ones being $3.39\,m^3$ and $5.66\,m^3$

2   A supply of oxygen gas in alloy steel cylinders charged to a pressure of 172.5 bars. Single cylinders of oxygen and D.A. or a bank of cylinders called a manifold may be used. A manifold supplying several blowpipes usually has an acetylene safety valve and two line pressure gauges. Copper pipe is never used for conveying acetylene due to an explosive compound being formed.

3   Pressure regulators for each gas to reduce the cylinder pressure to a suitable value for welding. (The regulator pressure screw should always be slackened off after welding has finished.)

4 Rubber canvas hose with special connections.

5 Blowpipe with set of nozzles (nozzle size may indicate the approximate consumption of the gas in litres/hour using a neutral flame).

6 Special tinted welding goggles.

7 A spark lighter.

Note: $1 \text{ bar} = 10^5 \text{Pa} = 10^5 \text{N/m}^2$.

Discharge Rates of Cylinders

The hourly withdrawal rate of gas from an acetylene cylinder must not be greater than 20% (one fifth) of its contents, otherwise acetone may be drawn off. (The acetylene cylinder should always be stood upright.)

Excessive withdrawal rates of oxygen, especially in cold weather, can cause "icing up" of regulators. Hot water should be used to thaw out frozen valves and regulators, NOT naked flames.

Gas	Ground Colour of Cylinder	Typical Use
Acetylene	MAROON	Welding, Cutting
Air	GREY	Brazing
Argon	BLUE	T.A.G.S., M.A.G.S., Pl.C.
Carbon dioxide	White strip Syphon type BLACK	M.A.G.S.
Helium	MED. BROWN	T.A.G.S., M.A.G.S.
Hydrogen	RED	Welding, Cutting
Methane	RED	Welding, Cutting
Nitrogen	DARK GREY (BLACK NECK)	T.A.G.S.
Oxygen	BLACK	Welding, Cutting

T.I.G. or T.A.G.S. = Tungsten arc gas shielded welding
M.I.G. or M.A.G.S. = Metal arc gas shielded welding
Pl.C. = Plasma cutting

IMPORTANT NOTE The new colour code to be introduced with the new ISO code for hoses will be:

BLUE—Oxygen
RED—Fuel Gases (except LPG)
BLACK—Air and Non-Combustible Gases
ORANGE—LPG (propane and butane)

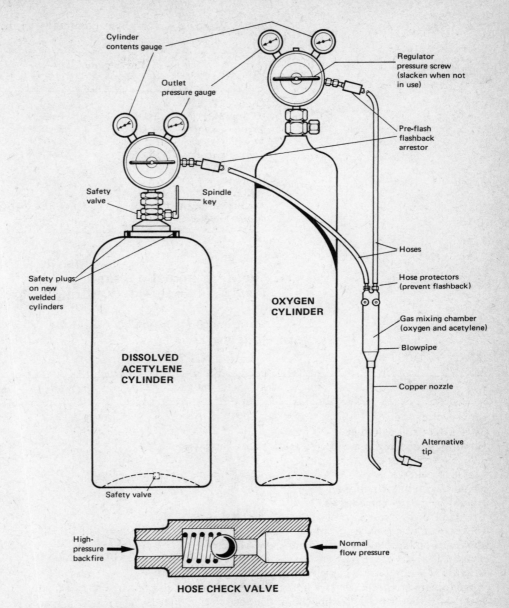

Cylinder contents gauge

Outlet pressure gauge

Regulator pressure screw (slacken when not in use)

Pre-flash flashback arrestor

Safety valve

Spindle key

Safety plugs on new welded cylinders

OXYGEN CYLINDER

Hoses

Hose protectors (prevent flashback)

Gas mixing chamber (oxygen and acetylene)

Blowpipe

Copper nozzle

Alternative tip

DISSOLVED ACETYLENE CYLINDER

Safety valve

High-pressure backfire

Normal flow pressure

HOSE CHECK VALVE

Fig. 6.1 Oxy-acetylene welding equipment

Some other gas mixtures in common use are as follows:

Gas	Ground Colour of Cylinder	Band Colour	Use
Argon + Carbon dioxide	BLUE	GREEN	M.A.G.S.
Argon + Oxygen	BLUE	BLACK	T.A.G.S., M.A.G.S.
Argon + Oxygen	BLUE	BLACK	M.A.G.S.
Carbon dioxide + Argon	BLUE	GREEN	M.A.G.S.
Hydrogen + Nitrogen	RED	GREY	PI.C.
Argon + Nitrogen	BLUE	GREY	PI.C.
Argon + Hydrogen	BLUE	RED	PI.C.

Welding Investigation 1
Oxy-Acetylene Welding Equipment

With information obtained from the process workshop, complete the Table.

Gas	Oxygen	Acetylene
Colour of cylinder.		
Pressure of full cylinder.		
Hand of thread of pressure regulator-where it screws into the cylinder.		
Metal of cylinder.		
Metal of regulator gauges.		
Height of cylinder in metres.		
Diameter of cylinder in metres.		
Differences of connecting nuts.		
Which direction should the cylinder key be turned to. open the valve?		
What is the number of cubic feet the cylinder which you have inspected holds when full? (There are various sizes.)		
Hand of thread of the welding hose connections on the regulator.		
What is the maximum reading that the outlet pressure gauge reads?		
What is the maximum reading that the contents pressure gauge reads?		
If you have a manifold in your college, list the number of cylinders actually coupled to the line.		
Is there a hose check valve fitted? State its location.		

QUESTIONS

1) From which metal is the blowpipe tip made?
2) Why is this metal used?
3) Why should cylinder valves be opened very slowly?
4) What would be the maximum withdrawal rate per hour of acetylene from a $3.39\,m^3$ cylinder?
5) What does the nozzle size indicate in your particular process workshop?
6) When a hose protector is fitted and the valves are open, the gas can flow:
a) from the blowpipe to the cylinder
b) from the cylinder to the blowpipe,
c) both ways,
d) not at all.
7) Can the cylinder contents gauge be altered by the pressure regulating screw?
8) Why must the pressure regulating screw be always screwed out before the spindle valve is opened?
9) What is the purpose of the regulators?

Welding Demonstration 1
Leftward Method of Welding a Butt Joint

AIM Demonstration of oxy-acetylene LEFTWARD welding of a butt joint in 1.6 mm low carbon steel and safe use of welding equipment.

NOTE TO LECTURER Both tacking and taper gap can be demonstrated.

THEORY This is a welding technique in which the flame is directed towards the unwelded part, and the filler rod, when used, is directed towards the welded part of the joint. See Fig. 6.2.

MATERIAL Two pieces low carbon steel, approx. $150 \times 80 \times 1.6$ mm.

PREPARATION Level, clean and straight.

EQUIPMENT Oxy-acetylene equipment including blowpipe and No. 3 nozzle, 1.6 mm filler rod, goggles, cylinder key, spark lighter, tongs, stop clock (see Table 6.1 on page 44).

PROCEDURE

Safety points and checks, lighting-up procedure, backfire, etc.
   Neutral flame, set clock and weld; stop clock; and note time.

QUESTIONS

Complete the following with the help of your lecturer:

1) Why are tinted lenses of ordinary spectacles no substitute for correct gas welding goggles?

2) What do the marks G.W.4 signify on the lens of gas welding goggles?

3) Why do you think it safer to pick up hot metal with tongs rather than with asbestos or leather gloves?

4) Give two sensible reasons why gas welders should not work very close to each other.

5) Contraction of the weld metal upon cooling causes the taper gap to close/open as the weld progresses.

6) Why do you think the demonstration you have just had would be dangerous if carried out in a confined space?

7) Why should blowpipes always be turned off when left unattended?

8) Name the position in which the weld was carried out.

9) What type of joint (see Table 6.1) is this?

10) Where should the spindle key be kept during welding?

11) State the nozzle size used as a number, and the quantity it passes in litres of gas per hour.

12) State the flame setting used.

13) State the steady regulator pressures during welding for both oxygen and acetylene.

14) What was the angle of SLOPE of the blowpipe?

15) What was the angle of TILT of the blowpipe?

16) Oil or grease is not used on high pressure oxygen connections because:

a) It may cause an explosion

b) It causes spanners to slip

c) It fouls up the connections

d) It leads to loosening of connections.

SAFETY

BACKFIRE This is when the flame flashes back up the nozzle and is arrested at the mixer or injector in the blowpipe body. Common causes are: incorrect pressure for tip, tip blocked by spatter, etc., overheating of nozzle, nozzle held too close to the work, trying to light before a free flow of both gases is obtained, and nozzle or connections not tight.

ACTION Promptly turn off the blowpipe valves, especially the acetylene to prevent soot forming in the blowpipe. A severe backfire is recognised by black smoke and sparks and a squealing noise issuing from the end of the nozzle. This may result in overheating, damage to the blowpipe, and even a flashback into the hoses. The cylinder valves should be closed promptly and hoses and equipment checked before welding is commenced. Overheated nozzles or tips may be cooled in a bucket of water.

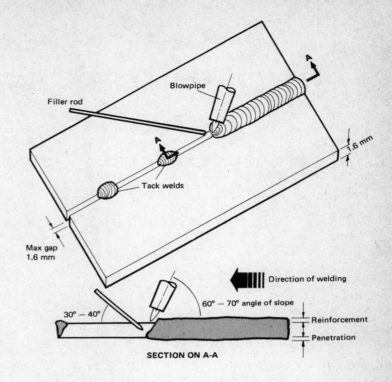

Fig. 6.2 Leftward welding
Note: Blowpipes must only have sufficient sideways movement to keep both edges melting

Blowpipe

Filler rod

1.6 mm

A

Tack welds

Max gap
1.6 mm

Direction of welding

$30° - 40°$

$60° - 70°$ angle of slope

Reinforcement

Penetration

SECTION ON A-A

7 Heat and Temperature Measurement

There is a fundamental difference between temperature and quantity of heat. An example which explains the difference would be the temperature of a candle flame and the temperature of a bucket of warm water. One burns the hand and the other does not. Yet despite the higher temperature of the candle it would take hours for it to warm the bucket of water, indicating that there is a greater *quantity* of heat in the water despite its lower temperature. Obviously the length of time of heating (quantity of heat) is more significant than the temperature of the heating medium.

When the temperature of a heated component is to be measured, it is important that we know whether an *approximate* or *precise* measurement is required and whether a com-

Table 6.1 Oxy-Acetylene Welding
Edge Preparation, Technique, Speed and Gas Consumption

(T) Thickness of Metals mm	Plate Gap (max) mm	Type of Edge Preparation	Technique	Name of Joint	Dia. of Weld Rod mm	Average Speed cm/h (complete)	Nozzle Size	Operating Pressure Oxy. bar	Operating Pressure Acet. bar	Gas Consumption Oxy. bar	Gas Consumption Acet. bar
0.9	nil		Leftward	Raised Edge Butt	—		1	0.14	0.14	28	28
1.2	½T			Open Square Butt	1.6 to 3.2		2	0.14	0.14	57	57
2.0	½T						3	0.14	0.14	86	86
2.6	½T	80—85°		Single Vee Butt			5	0.14	0.14	140	140
3.2	½T						7	0.14	0.14	200	200
4.0	3.2		Rightward	Open Square Butt			10	0.21	0.21	280	280
5.0	3.2						13	0.28	0.28	370	370
6.5	3.2						18	0.28	0.28	520	520
8.2	1.6	80—85°		Single Vee Butt			25	0.42	0.42	710	710
10.0	3.2						35	0.63	0.63	1000	1000
Lap Fillet Welds											
1.6	nil		Leftward	Lap Fillet	1.6		3	0.14	0.14	86	86
3.2			Rightward	Lap Fillet			7	0.14	0.14	200	200
5.0			Leftward	Lap Fillet	3.2		10	0.21	0.21	280	280
8.2			Rightward	Lap Fillet			18	0.28	0.28	520	520
Tee Fillet Welds											
1.6	nil		Leftward	Closed Tee Fillet	1.6 to 3.2		3	0.14	0.14	86	86
3.2			Rightward				7	0.14	0.14	200	200
5.0			Leftward				13	0.28	0.28	370	370
8.2			Rightward				25	0.42	0.42	710	710
Corner Joints (Fillet)											
1.6	½T		Leftward	Closed Corner	1.6 to 3.2		3	0.14	0.14	86	86
3.2			Rightward	Open Corner			7	0.14	0.14	200	200
5.0							18	0.28	0.28	520	520

paratively low or high temperature is to be measured. The following is a selection of the more common methods used in fabrication and welding.

1 Colour Change of Metals (Approx.)

Steel when heated passes through noticeable colour changes as its temperature alters and this enables an approximate temperature to be estimated. (See Fig. 7.1.) Other metals such as copper and its alloys only show a dull red to light orange. Cast iron becomes a dark red, but aluminium has no noticeable colour change.

Applications Forging. Quenching. Pre-heat.

2 Heat-sensitive Indicating Paints and Crayons (Approx.)

These work on the principle that mixtures of mineral oxides and pigments will change colour or melt on heating. The surface should be free from rust, dirt and grease. The paint should be applied by brush or alternatively a "tempilstick" crayon may be used which is simply stroked on the job and *melts* at the stated temperature. They range in calibrated steps from 30°C to 1600°C.

Paints and crayons are used to estimate the temperature of the actual component and not the surrounding temperature. They must not be heated directly as the heat from the metal should cause the colour change. There is the reversible type which, for example, changes from red to black at 70°C and upon cooling reverts to its original colour, red, the change being instantaneous. These are used normally on non-metals. Metallic surfaces should be cleaned and lacquered to avoid contact between paint and metals when this type is used on metal.

The non-reversible type changes colour, e.g. red to green at 800°C and remains that colour regardless.

The "tempilsticks" are marked with the change temperature. Colour changes are indicated for paints. They range in temperature from 80 to 1400°C.

Applications Pre-heat heat treatment.

An extension of this principle is the use of white soap or matchsticks used to judge the pre-heat temperature of aluminium. The aluminium is rubbed with either the soap or matchstick and the temperature estimated from the resulting brown mark. Light brown approx 300°C, dark brown 400°C.

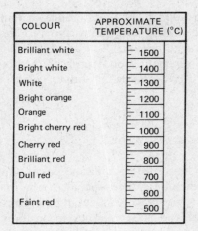

COLOUR	APPROXIMATE TEMPERATURE (°C)
Brilliant white	1500
Bright white	1400
White	1300
Bright orange	1200
Orange	1100
Bright cherry red	1000
Cherry red	900
Brilliant red	800
Dull red	700
	600
Faint red	500

Fig. 7.1

Fig. 7.2

3 Heat Cones or Pyramids (Seger type) (Approx.)

These are pyramids of mixtures of elements such as Kaolin
Lime, etc., varied to give a range of melting temperatures so
that the top of the pyramid arches over to touch its base,
indicating that the temperature has been reached (Fig. 7.2).
Three are usually used with approx. 20°C difference in temperature between each, the middle one being the required
temperature. Temperature range between 400 and 2000°C.
Application Used in fabrication for furnace heating
aluminium and magnesium alloys. Kilns for bricks, pottery,
etc.

4 Indicating Pellets and Dots (Approx.)

A similar material of manufacture as crayons and paints. The
pellets and dots melt immediately at their specified temperature with an error of not more than ±1% at temperatures
ranging from 45°C to 900°C. The dots are stuck in position.
Applications Assessing the temperature distribution of tur-
bine blades under test conditions.

5 Liquid Expansion Thermometers (Precise)

These thermometers of the mercury-in-glass or mercury-in-
steel type work on the principle that the heated mercury
expands up a fine bore capillary tube calibrated in °C. With the
mercury-in-steel type, the capillary tube is flexible and connected to a Bourdon-type gauge. All air is exhausted from the
capillary tubes at manufacture and the ends sealed. The
mercury-in-glass measure up to 340°C and in steel up to 395°C.
Applications Temperature of liquids, mounting resins, fats,
room and workshop temperatures.

6 Thermocouple Pyrometer (Precise)

If two different metals such as platinum and platinum/rhodium
are joined together (surrounded by a refractory sheath for
protection) and heated at the joint, and the other end kept
cold, then a voltage is set up in the circuit. This is called an
electromotive force (e.m.f.). A millivolt meter which is calibrated to read °C is connected into this electrical circuit which
will measure the e.m.f.. The value of the e.m.f. is proportional
to the temperature of the joint. The temperature may be read
directly from the meter. The temperature range depends upon
the type of dissimilar metals used but temperatures up to
1600°C may be measured.
Applications Furnace temperatures for annealing, normalising and stress relieving. Molten metals.

Fig. 7.3 Thermocouple pyrometer

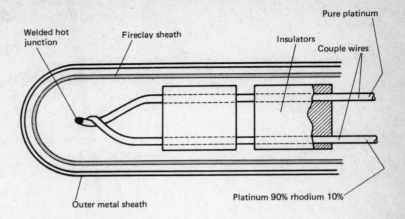

Welded hot junction
Fireclay sheath
Insulators
Pure platinum
Couple wires
Outer metal sheath
Platinum 90% rhodium 10%

7 Optical Pyrometer (disappearing filament type) (Precise)

This type compares the intensity of light radiated from the heat source with a known heat source.

The most common type is the disappearing filament type. An electric bulb filament is sighted against the hot body being measured. Current is passed through the filament causing light to be emitted. The current is adjusted by means of a variable resistor until the colour of the hot body and the filament colour just match, causing the filament to disappear. An ammeter measures the current used by the filament and is calibrated so that the temperature may be read directly. Flames and smoke cause variable readings. Temperatures in the range 700 to 1800°C may be measured.

Applications Furnace roofs and walls for collapsible hot spots. Molten metals. Heat treatment.

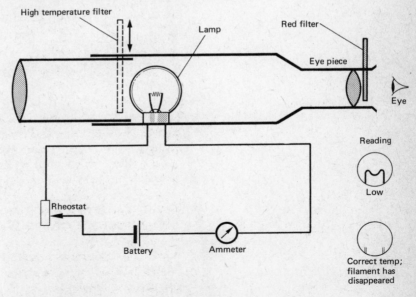

High temperature filter
Lamp
Red filter
Eye piece
Eye
Reading
Low
Rheostat
Battery
Ammeter
Correct temp; filament has disappeared

Fig. 7.4 Optical pyrometer

Materials Investigation 4

Cut two 150 mm square thin sheets, one of aluminium and one steel. Draw a cross from the diagonals with soap and dot these at regular intervals of 12 mm from the centre with indicating paints and crayons. Heat in the centre a spot, 1000°C for the steel and 600°C for the aluminium, using a neutral oxy-acetylene flame. Allow to cool. Note the colour changes of the soap on the aluminium and the oxide changes on the steel and compare these with the relevant temperature and colour changes of the paints.

8   The Oxy-Acetylene Flame

Acetylene is composed of Hydrogen and Carbon, as are most fuel gases. It is mainly the carbon which provides the intense heat and very high flame temperature (3100°C) when burned with oxygen. If sufficient oxygen is not provided, then the carbon is given off into the air as black, sooty smuts.

Acetylene has a very high proportion of carbon in it and if the oxygen is turned down to provide a flame with excess carbon, the carbon is taken into the steel to provide a high carbon surface, used for hard surfacing operations.

Fig. 8.1 The oxy-acetylene flame, showing the various zones

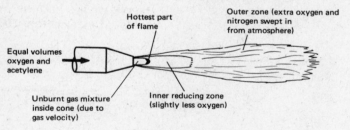

A neutral oxy-acetylene flame burns equal proportions of oxygen and acetylene and is reducing in nature, thereby reducing any iron oxide to iron and taking up the oxygen; consequently there is no need to use a flux when welding steel. It should be noted that iron oxide is not refractory.

Welding Demonstration 2
The Structure of the Oxy-Acetylene Flame

AIM   To demonstrate the three oxy-acetylene flame settings.
EQUIPMENT   Oxy-acetylene welding equipment, including goggles and flint lighter.

THEORY There are three distinct flame settings:

Fig. 8.2 Neutral flame
Cone tip hottest part approx. 3100°C

a) **Neutral Flame** (Fig. 8.1 and Fig. 8.2)
This flame burns equal quantities of oxygen and acetylene. (In practice, it is advisable to have the slightest possible acetylene haze at the cone tip to begin with.)

Fig. 8.3 Carburising flame

b) **Carburising Flame** (Fig. 8.3)
This flame has an excess of acetylene which results in a carbon-rich zone extending around and beyond the cone.
Note: Both the Neutral and Carburising flames are reducing in nature.

Fig. 8.4 Oxidising flame

c) **Oxidising Flame** (Fig. 8.4)
This flame has an excess of oxygen which results in an oxygen-rich zone just beyond the cone. This flame is obtained by setting to neutral and then turning the fuel gas down.

QUESTIONS
During the demonstration, note the answers to the following questions:

1) Which is the noisiest flame.
2) The longest flame.
2) The whitest flame.
4) The flame with the smallest cone.
5) The flame with the most pointed cone.
6) The flame with the inner feather.
7) What kind of gas is inside each cone?

Applications of the Various Flame Settings

Neutral Flame
This flame is used for the welding of steel, cast iron, stainless steel, copper and aluminium.

Carburising Flame

This flame is used for hard surfacing with hard surfacing rods. The parent metal is heated and the flame gives up carbon to the parent metal so lowering its melting point and allowing the rod to be deposited quickly without deep penetration. A very slightly carburising flame is often used for non-ferrous metals where the smallest amount of oxygen would be undesirable.

Oxidising Flame

Undesirable when very strongly oxidising, except for welding brass. When slightly oxidising it is used when brazing zinc coated sheet.

Welding Investigation 2
Effect on Metal Surface of Varying the Proportions of Oxygen and Acetylene

AIM   To note the effect on the surface of a metal when the proportion of oxygen and acetylene is varied.

THEORY   If oxygen is added to a metal it forms an oxide of the metal and the metal is said to have been OXIDISED.

If the oxide is then subjected to a REDUCING atmosphere in the form of a reducing flame, the oxygen is taken away and the oxide is reduced to the metal again.

EQUIPMENT   Oxy-acetylene welding equipment, clean copper sheet, approximately $60 \, mm \times 60 \, mm \times 20$ s.w.g. or suitable, no. 3 nozzle, emery cloth.

PROCEDURE

1) Light the blowpipe and adjust to a neutral flame, then heat the copper to a dull red in the centre. Remove the flame and allow to cool, noting the formation of oxide film.
2) Remove the blowpipe and reset to a carburising flame, and apply to the centre of the copper which will again have acquired an oxidised surface. Note the increase in area affected.
3) Remove the blowpipe and reset to oxidising condition. Apply again to centre of copper and note the immediate effect on the copper and also the effect of leaving the flame on the plate for a long time.
4) Now repeat (1) and (2) above and note the results.

QUESTIONS

1) Name the two flames which are reducing in nature
2) Where does the oxygen come from to form the oxide on the plate when the neutral flame is removed?
3) What happens to the thin film of oxide when the neutral flame and carburising flame are applied?

4) Why do you think the area affected by the carburising flame is greater?

5) What was the effect of prolonged application of the oxidising flame?

Welding Investigation 3
Effect of Flame Setting on Low Carbon Steel

AIM To note the effect on the surface of low carbon steel when the various flame settings are used.

EQUIPMENT Oxy-acetylene welding equipment, six pieces of low carbon steel, 50 mm × 50 mm × 1.6 mm each with a line scribed on centre, no. 3 nozzle, 1.6 mm dia. L.C. filler.

PROCEDURE 1
Stamp two of the plates on diagonal corners with N, two with O and two with C for later identification. Check that the equipment is in correct working order as demonstrated by the lecturer previously. Check that the pressure regulator control is NOT screwed in, then open the valve slowly.

Consult Table 6.1 (page 44) for correct pressures. Light the blowpipe and adjust to a Neutral Flame. Position the flame so that the end of the cone is not quite touching the plate marked N (the blue cone must not touch), and when the metal begins to melt and form a puddle, move the blowpipe slowly down the plate to form a fused surface bead, using the Leftward Technique.

Repeat the procedure using an Oxidising Flame on the plate marked O and a Carburising Flame on the plate marked C. Finally, using the three flame settings, see how long it takes for each to melt through the other plates near the edge after completing beads with filler.

Draw and complete the table and sketch each of the following flame settings: neutral, oxidising, carburising.

Specimen	Time to Penetrate Plate (secs)	No. of Reversals	File Hardness of Filler	Bead Appearance
Neutral Flame	N			
Oxidising Flame	O			
Carburising Flame	C			
N+Filler	NR			
O+Filler	OR			
C+Filler	CR			

PROCEDURE 2
Clamp each specimen in the vice as shown in Fig. 8.5 and using mole grips, reverse bend until it breaks (note the number of reversals) to test the ductility. File each of the beads made with the filler and record Hard, Very Hard or Normal.

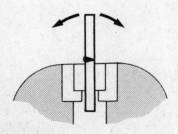

Fig. 8.5

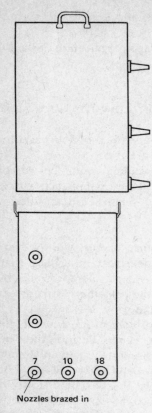

Fig. 8.6 Multi-nozzle tank

7 10 18

Nozzles brazed in

QUESTIONS
1) Which flame setting would you use for welding low carbon steel?
2) Which flame gave the bead with the best appearance?
3) Which flame gave the bead that was hardest to file?
4) Which flame gave the bead that flowed pasty with lots of sparks?
5) Add any other differences you noted.

Welding Demonstration 3 and Investigation 4
Gas Velocity, Nozzle and Filler Rod Size

THEORY For every thickness and size of plate there is an optimum-size nozzle and filler rod (if necessary). By adjusting the flame power and varying the welding technique, a variety of thicknesses and sizes may be welded using the same nozzle and filler rod, but only within certain limits. To vary the flame power, the gas velocity (which is the pressure or speed of the gas issuing from the nozzle) is adjusted by the oxygen and acetylene valves.

When the gas velocity becomes too high for the nozzle size, the flame snaps out, and when the gas velocity is too low backfiring results (the flame burns back into the nozzle). The nozzle size indicates the approximate consumption of each gas in litres per hour when using a neutral flame.

DEMONSTRATION To show the effect of pressure and nozzle size on velocity.

EQUIPMENT 1 Multi-nozzle tank (Fig. 8.6), plus corks, three 300 ml beakers, stop clock.

PROCEDURE
1) Put corks on end of nozzles and fill tank with water.
2) Remove corks on the three lower nozzles and note the distance reached by the water for each nozzle (all same water level and pressure).
3) Place a beaker under each jet and note the time taken for first to fill to 300 ml mark. Cork all nozzles and note volume in all beakers for same time.

Nozzle Size	Time (secs)	Volume (ml)
7		
10		
18		

Nozzle Size	Position	Pressure	Distance (cm)
7	Top	Low	
7	Middle	Medium	
7	Bottom	High	

4) Now uncork all no. 7 nozzles (different levels and water pressure) and note distance reached by each jet. Then complete the Tables. (Note: draw the water curve.)

EQUIPMENT 2 Blowpipe, nozzles 7, 10, 18 and balanced funnel. Air connection only (Fig. 8.7).

PROCEDURE
1) With blowpipe clamped to board and using air pressure of 0.14 bar fit varying nozzle sizes 7, 10, 18 and mark where each is pushed by gas velocity.
2) Note what happens when the gas pressure on the 7 nozzle is steadily increased from zero to 0.63 bar and complete the Table.

Nozzle Size	Gas Pressure	Velocity Mark
5	0.14 bar	
10	0.14 bar	
18	0.14 bar	
7	0.14 bar	
7	0.42 bar	
7	0.63 bar	

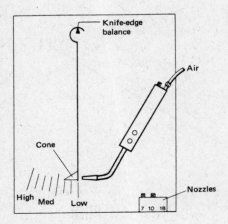

Fig. 8.7

3) Now connect and light the blowpipe and, using high, medium and low gas pressures with equal oxygen and acetylene pressures, judge the correct gas velocity for the nozzle size.

INVESTIGATION Involving filler rod size and gas velocity.
EQUIPMENT Ten pieces of low carbon steel, 100 mm × 50 mm × 3 mm. No. 7 nozzle, gas welding equipment and goggles, etc., tongs. (See Table 6.1, page 44.)

PROCEDURE
Tack weld each pair in the close corner position at 90° (Fig. 8.8) and using a neutral flame and the Leftward Technique, weld as indicated in the Table below. Inspect the weld profile, then flatten the joint to 180°. Complete the Table.

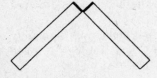

Fig. 8.8

Joint	Gas Velocity Pressure	Size Filler	Profile	Flatten
1 corner	High	Nil		
2 corner	Low	Nil		
3 corner	Medium	Nil		
4 corner	Correct	1.6 mm		
5 corner	Correct	2.4 mm		
6 corner	Correct	3.2 mm		

Fig. 8.9 Rightward welding

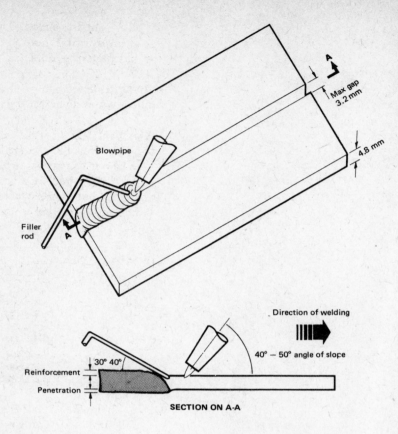

Welding Demonstration 4
Rightward Method of Welding a Butt Joint

AIM Demonstration of the Rightward method of welding a butt joint in 4.8 mm low carbon steel.

THEORY This is a welding technique in which the flame is directed towards the welded part, and the filler rod is directed toward the unwelded part.

MATERIAL Two pieces of low carbon steel, approximately 150 mm × 80 mm × 4.8 mm. Filler, etc. (See Table 6.1, page 44.)

PREPARATION Level with edge preparation filed square and clean.

EQUIPMENT As for Leftward demonstration, but 13 nozzle and 2.4 mm filler.

PROCEDURE
As for Leftward demonstration.

QUESTIONS
1) State the time taken to complete the weld.
2) What is the type of joint called?

3) State the nozzle size used and the litres of gas per hour it passes.
4) What flame setting was used?
5) What was the steady regulator pressure in bars for both oxygen and acetylene?
6) What was the angle of slope of the blowpipe?
7) What was the angle of tilt of the blowpipe?
8) State the position of welding.
9) Show, by simple sketch, the movement of the filler rod.
10) Show the movement of the blowpipe.

Welding Demonstration 5
Comparison of Leftward and Rightward Welding

AIM To compare Leftward and Rightward Welding Techniques.

THEORY In general it can be said that

Leftward welding is carried out on **thin gauge sheet and pipe** and

Rightward welding on **thicker plate and pipe**, with sheet or wall thicknesses of 4.8 mm being the changeover line.

EQUIPMENT Oxy-acetylene welding plant, as for L and R.

Two plates, 150 mm × 80 mm × 4.8 mm square edge.

Two plates, 150 mm × 80 mm × 4.8 mm bevelled as Table 6.1 on long side.

PROCEDURE 1
1) Leftward weld together the two bevelled plates.
2) Rightward weld together the two unbevelled plates.
3) Carry out a reverse bend on a 40 mm strip (see page 56).

QUESTIONS
1) Time taken for leftward method?
Time taken for rightward method?
2) Which technique uses the most welding rod?
3) Which technique uses the most gas?
4) Which technique causes the most warping of the plates?
5) Which technique allows the best view of the weld pool during welding?
6) Which technique gives the best surface finish?
7) Which technique produces the most uniform penetration?
8) After bending, was any cracking observed? If so, which specimen?
9) Which technique gives a beneficial heat treatment to the weld during welding?
10) Give a typical industrial use of (*a*) the leftward technique, (*b*) the rightward technique.
11) Which technique do you think is the easier to perform?
12) Which oxy-acetylene technique would you advise for:
a) 2 metres of welding on 20 s.w.g.
b) 10 mm thick pipe wall, 150 mm diameter.

PROCEDURE 2

Now weld the following joints: LAP, TEE FILLET and CORNER, using the Leftward Technique and then the Rightward Technique and compare the results.

Use information from Table 6.1 using light to heavy gauge plate. Summarize the advantages of each process.

9 Bend Tests

A bend test may be a *free bend* test where mole grips, bending bars, pliers, etc., are used to bend the sheet or plate without using a former (Fig. 9.1); or a *guided bend* test (Fig. 9.2), where the plate is guided between rollers and the force on the former is supplied by a hydraulic ram or mechanical means (9.2) or bent over a former (9.1). The purpose of the bend test is to assess the soundness of the weld metal and the ductility of the weld joint, especially where the parent metal and filler metal have fused together. In other words, the efficiency of the welded joint. (See chapter on Inspection, page 58, on the use of macro-etching prior to bending.)

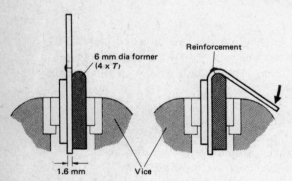

Fig. 9.1 Free bend
For sheet 1.6 mm thick, reinforcement is to be removed for metal arc welds but left on for oxy-acetylene welds

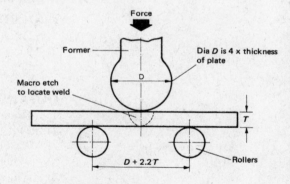

Fig. 9.2 Guided bend
The sharp corners at the bend must be rounded off
T = plate thickness

Cupping Test

A common workshop method of testing butt welds in thin sheet is the cupping test, carried out with the weld over a recessed block or pipe and repeatedly struck with a ball pein hammer to form a hemispherical indentation. If there is lack of fusion or other serious defects present, then the material will split. If no cracks appear, then it shows that the joint is ductile, malleable and strong. In addition to thin sheet steel, other metals such as copper and aluminium are tested by this method. The depth of indentation should be about one third the diameter.

The above test must not be confused with the Erichsen cupping test performed in the laboratory. This is to test the metal for its suitability for deep drawing operations. A hemispherical punch forces the sheet through a collar and, depending upon the depth of indentation and any signs of cracks, an opinion may be formed regarding its deep drawing and pressing suitability.

QUESTIONS

1) What does the .2T allow for in a) plate thickness and b) between the former D and the rollers? (See Fig. 9.2.)

2) The plate edge is given a macro-etch prior to bending mainly to enable
a) the root to be seen;
b) the defects to be seen;
c) the weld to be located;
d) the penetration to be seen.

3) The centre line of the former should be in line with the centre of the weld to ensure:
a) the weld joint is fully tested;
b) the parent metal is tested;
c) the plates are equal either side of the weld;
d) both weld and parent metals are tested.

4) Why is it dangerous to try and bend thin sheet using only your hands to test the weld?

5) The reason that the former D is made four times the thickness of the plate is to enable:
a) the maximum stress to be exerted in the weld;
b) a radius to be formed which adequately tests the weld;
c) a large radius to be formed;
d) a good clearance to be used.

6) During bending, on the outside of the bend the metal is:
a) buckled; b) stretched; c) twisted;
and on the inside, the metal is:
a) flattened; b) warped; c) compressed.

7) If the weld metal during bending:
a) gets longer; b) fractures; c) becomes thinner;
it is not a ductile weld.

8) Make a sketch, in good proportion, to show a workshop cupping test. State a) when the test would be used; b) what the test reveals.

10 Inspection: Macro-Etch and Micro-Etch

Suggested Method for Low Carbon Steel

a) Preparation of Surface for Etching

The surface should be filed with a coarse file until all deep marks are removed. It should then be filed at right angles to the original coarse file marks with a smooth file, then emery paper, e.g. M, FM, 0, 00, the direction of polishing being at right angles to the marks made by the previous paper in each case. Polishing should be continued until the scratches of the previous paper have been removed before proceeding to the next finer grade. This procedure shows the means by which a first-class finish may be obtained.

For micro-etching, the specimen is finished on a polishing wheel with alumina or diamond paste.

b) Etching for Macro-Examination

In general for steel an 0 emery finish will be smooth enough for a satisfactory etch to be obtained for macro-examination. A suitable etching solution is as follows:

> 10–15 ml nitric acid (sp. gr. 1.42)
> 90 ml alcohol (industrial spirit)

c) Etching for Micro-Examination

For cast iron and steel, 2–3% solution of nitric acid in alcohol.

The etching is carried out either by swabbing the surface with cotton wool or by immersing the specimen in the etching solution until a good definition of the structure is obtained. The specimen should then be washed in water, preferably hot, followed by rinsing with acetone or alcohol (industrial spirit) and dried in a current of air. If specimens are to be preserved, a coating of thin oil or colourless lacquer is of advantage.

The naked eye or a low powered glass (×10) is used after the macro-etch.

Macro-etching of steels is carried out for various reasons, the main ones being as follows:

1 To show the extent of the fusion boundary (penetration).

2 To locate the weld (for bend tests).

3 To indicate certain defects such as cold laps, lack of fusion and penetration (porosity, slag inclusions and cracks may also be seen although these are not etched).

4 To show the columnar crystal structure of the deposited weld.

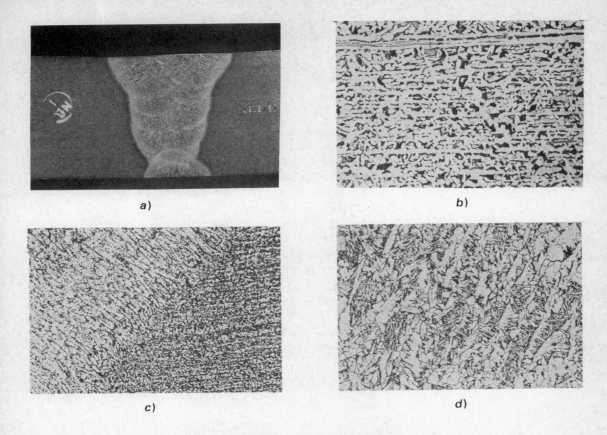

a)

b)

c)

d)

Typical examples of weld macro and micro structures in low carbon steel

a) shows the macro section of a multi-run butt weld. (Etchant 10% Nital.) Note the identification stamps, columnar grains, recrystallisation, and order of the runs.

b) shows a micro structure (×90) of the parent plate material. (Etchant 3% Nital.) Note the rolling direction of the plate.

c) shows the weld junction (×90). (Etchant 3% Nital.) Note on the left the change of crystal growth direction.

d) shows the columnar structure of the weld (×90). (Etchant 3% Nital.) Note the steep angle of columnar crystal growth.

Micro-examination of steel is carried out under a microscope at between 100 and 500 diameters to show grain size and shape, grain structure (ferrite, pearlite, etc.), and micro defects such as cracks and porosity.

Metal	Reagent	
18/8 Austenitic Stainless	Hydrochloric acid	15 cc
	Nitric acid	5 cc
	Water	100 cc
Aluminium and Alloys	Hydrofluoric acid	10 cc
(Keep hydrofluoric acid in	Nitric acid	1 cc
plastic bottle and beware	Water	200 cc
of dangerous fumes)	(Remove black film	
	in dilute nitric)	
Copper and Alloys	Nitric acid	25 to
	Water remainder	45 cc

11 Weld Assessment (Visual Inspection)

Not necessarily in order of importance.

1 Dimensions of the Weld Deposit
Measured by the LEG LENGTH.

2 Shape of Profile for Fillet Welds
Welds should be uniform and slightly concave, or convex, or if possible a true mitre.

3 Uniformity of Surface
The height and spacing of the ripples should be uniform and the width of the bead constant.

4 Degrees of Undercut
The welded joint should be free from undercut, but small amounts may be disregarded.

5 Smoothness of Joints (where the welding has recommenced)
These should be fairly smooth without hump or crater.

6 Freedom from Surface Imperfections
There should be no pin-holes, cavities, cracks, craters or burnt-on scale.

Inspect Fig. 11.1 and
1) State the shape of the profile.
2) State the type of joint.
3) State the main defect.

Fig. 11.1

12 Weld Defects

Fig. 12.1

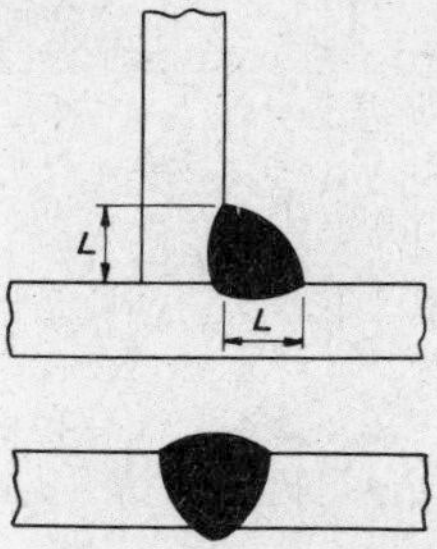

A weld defect is an imperfection in the weld which may eventually lead to failure of the weld joint under the service conditions for which it is designed.

These imperfections may be detected by:
a) Visual inspection of the surface;
b) Cutting the joint and preparing for macro-inspection.
c) Mechanical testing (Bend Tests in this case).
Other methods are used in addition to the above.

A **Satisfactory Profile** is shown in Fig. 12.1.

Fig. 12.2

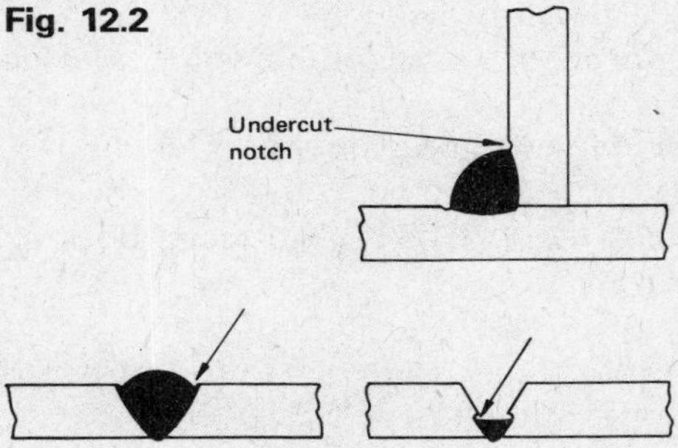

Undercut notch

Undercut (Fig. 12.2)

A groove or depression parallel and adjacent to the sides of the weld, resulting in thinning of parent metal at the toe of the weld
Causes
a) A too rapid rate of travel of blowpipe or electrode.
b) Excess heat build-up.
c) Incorrect angles of electrode or blowpipe.

Shape of Profile

The profile should be
  uniform and slightly convex on butt welds
  slightly concave or convex for fillet welds,
with a smooth join up of toes to parent metal and a completely filled groove.

Fig. 12.3

Incompletely filled groove

1) *Incompletely Filled Groove* (Fig. 12.3).
Causes
a) A too rapid rate of travel with electrode or blowpipe.
b) Too small an electrode or rod.

Fig. 12.4

2) *Overfilled Groove* (Fig. 12.4).
Causes
a) Welding current or blowpipe size too small.
b) A too slow rate of travel.
c) Electrode or rod too large.

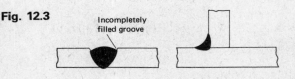

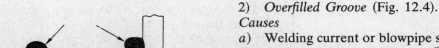

Overlap (Fig. 12.5)

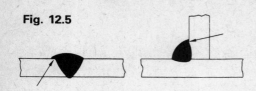

Fig. 12.5

Excess weld metal at the toe of a weld covering the parent surface but not fused to it.
Causes
a) Welding current too low and travel too fast of electrode or blowpipe.
b) Melting of filler rod before parent metal.

Penetration

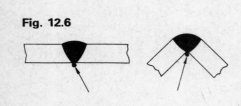

Fig. 12.6

1) *Excessive* (Fig. 12.6): This is where excess weld metal is protruding through the root of the joint. The penetration bead should not normally exceed 1.6 mm especially on pipes.
Causes
a) Too high welding current.
b) Concentration of heat.
c) Slow speed of travel of blowpipe or electrode.
d) Unsuitable edge preparation.

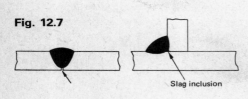

Fig. 12.7

Slag inclusion

2) *Incomplete* (Fig. 12.7): Weld metal does not extend to root.
Causes
a) Too rapid a speed of travel of electrode or blowpipe.
b) Insufficient current or flame power.
c) Melting of filler without fusing parent metal.

Slag Inclusions (root) (Fig. 12.7)

Causes
a) Current too low, electrode too large.
b) On fillets, incorrect angles, insufficient cleaning on multi-runs.
c) Speed of weld too fast.

Lack of Fusion (Incomplete Fusion)

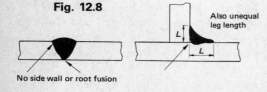

Fig. 12.8

Also unequal leg length

No side wall or root fusion

Lack of union between weld metal and parent metal (Fig. 12.8).
Causes
a) Welding current too low.
b) Insufficient heat.
c) Too rapid travel with blowpipe or electrode.
d) Inadequate preparation.
e) Dirty surface.
f) Incorrect angle of blowpipe or electrode.
g) Incorrect arc length.
h) Melting of filler before plate.

Fig. 12.9

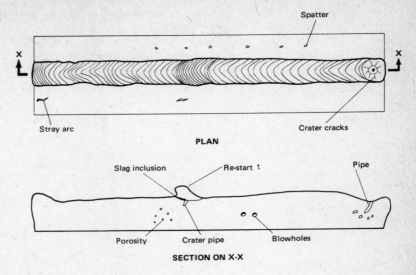

PLAN

SECTION ON X-X

Surface Defects

These could be in the form of blowholes, porosity, slag traps, burnt-on scale, cavities, bad "join ups" showing humps or craters with cracks, uneven bead width, stray arcing and burnt-on spatter, extremely coarse ripples, unequal fillet weld leg length. See Fig. 12.9.

Causes
a) *Blowholes:* 1.6 mm plus in dia. caused by gas from moisture and dirty plate. Also long arc length.
b) *Porosity:* Less than 1.6 mm dia. group of gas pores. Damp electrodes, rust, moisture and grease on plate.
c) *Uneven Bead:* Uneven rate of travel and fluctuating arc lengths.
d) *Stray Arcing:* Small craters causing high stress points and hardness. Use a striking plate.
e) *Coarse Ripples:* Erratic movement of electrode and blow-pipe.
f) *Cracks:* Small tack welds, long arc length, incorrect electrode and rod.
g) *Crater Pipe:* Depression due to shrinkage.
h) *Unequal Leg Length:* Incorrect angle of rod and blowpipe or electrode. Too much heat build-up.

EXERCISE
Using the test weld beam samples and joints shown, which you sectioned during investigation of factors relating to the metal arc and oxy-acetylene processes, sketch as in the table (p. 64) the weld bead cross-section, and name the defects and state the probable cause of the defects.

| Type of Joint | Visual Inspection | Bend Tests | | Macro with Defects | Process and Technique | Probable Causes |
		Normal	Rev			
Tee Fillet	Spatter Slight undercut.	Nil	Nil	6 mm	Metal arc	Welding too fast. Angle of slope and tilt wrong.
Open Square Butt	Uneven bead. No penetration.	Good 180°	Broke at 90°	1.6 mm	Oxy-acetylene, leftward	Insufficient tacks. Flame power low.
Lap						
Closed corner						

13 The Electric Arc and the Functions of the Electrode Coating

When the welding circuit is "closed" by striking the electrode on to the work and withdrawing it slightly, an arc is formed. This contact is to enable a flow of current in the form of electrons to take place after the initially high voltage has overcome the resistance to current flow (sometimes called ionization of the arc gap).

The arc causes the parent metal to melt or fuse.

The metal core of the electrode conveys the electrical energy to the arc and is melted along with the flux coating to form molten droplets of metal and flux. The arc is now composed of regions of very high temperature gases (approximately 6000°C) mainly obtained from the flux coating. The force of the arc, helped by gravity and surface tension, projects the molten droplets into the weld pool where they solidify under the protective covering of the solidified flux, now called slag. The flux also provides a shield of gas which protects the molten metal both at the electrode tip and the molten weld pool.

In addition the flux supplies salts which provide ionised particles to assist re-ignition of the a.c. arc.

The functions of the electrode coating may be summarised as follows:

1 To provide a protective shield of gas around the arc and the molten metal and so prevent oxygen and nitrogen from the atmosphere entering the weld metal.

2 To make the arc stable and easy to control ("smooth" arc).

3 To replenish any deficiency caused by oxidation of certain elements from the weld pool during welding and to ensure the weld has satisfactory mechanical properties.

4 To provide a protective slag which also reduces the cooling rate of the weld metal and thus reduces brittleness due to chilling.

5 To help to control (along with the welding current) the size and frequency of the droplets of weld metal.

6 To enable different positions to be used.

14 Basic Functions of Metal Arc Welding Equipment and its Safe Use

Theory

Basically a metal arc welding plant consists of the following important items:

a) The power source, which may be alternating current (a.c.) or direct current (d.c.).

b) The welding lead cable and electrode holder.

c) The welding return cable (NOT earth lead) and clamp.

d) The welding earth.

Two types of welding current are used: a.c. which changes from negative to positive at the frequency of the supply; and d.c. which flows in one direction only.

The a.c. "mains" supply is not suitable for welding because the voltage is too high and the current too low. A transformer is used to change this to suitable values for welding. i.e. low voltage, high amperage but still alternating current.

An alternative to this method is to use the a.c. "mains" power supply and convert this from a.c. to d.c. by using what is

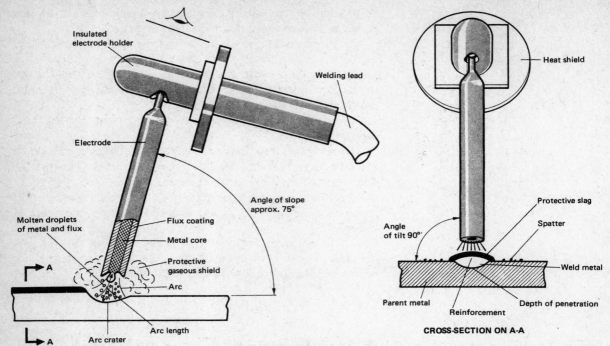

Fig. 14.1 Diagrammatic view of bead of weld on flat plate by manual metal arc process

termed a Rectifier (still low voltage, high amperage for welding).

A generator can supply d.c. either from a fuel-driven motor or from current taken from the mains to drive the motor. There are other forms of supply also.

Open Circuit Voltage (O.C.V.) and Arc Voltage

The open circuit voltage is the voltage available at the output terminals of a welding set ready to weld but carrying no current and is usually limited to between 80 to 100 volts for a.c. and 60 to 80 volts for d.c. This initially high voltage enables the arc to be struck. As soon as the arc is struck the voltage between the electrode tip and the work drops to less than half the O.C.V. This voltage is now termed **Arc Voltage** and varies also with the arc length and the type of electrode being used.

When an arc is struck, the current flows along the welding cable to the insulated holder, down through the electrode, and back along the return cable to the welding plant. If this circuit is broken at any time, current stops flowing. There should be an earth bonded on to the plant and one as near to the work as possible to protect the operator who must at no time become part of the circuit.

Safety

1   ALWAYS: wear a recommended face shield complete with coloured glass (which filters out the dangerous infra-red and ultra-violet rays and prevents Arc Eye). Plain glass protects the coloured glass.
REMEMBER—Goggles can be bought, eyes cannot! There are no jobs for blind welders.
2   Protect exposed skin from burns caused by the rays and hot metal.
3   Wear leather gloves.
4   Wear a leather apron.
5   Do not pick hot metal up with a gloved hand, use tongs.
6   It is dangerous to weld parts which have traces of the degreasing agent trichloroethylene still on them and harmful to breathe zinc fumes off galvanized plate.
7   Do not weld standing on wet concrete.
8   Ensure that all current-carrying cables and connections are insulated and do not become a hazard to others.
9   Ensure no inflammable material is in the near vicinity of the welding operation where it could be ignited, i.e. oil, rags, paper or waste.
10   When working in confined metal tanks, damp conditions or at a height using a.c. equipment, it is recommended that a low voltage safety device is fitted to the output side of the power source to prevent severe electric shock.

Fresh air during welding is essential, either natural or ventilated.

Welding Investigation 5
Metal Arc Welding

Fill in the missing words on the diagram (Fig. 14.2) with the help of your lecturer, then answer the questions.
1   What is the "mains" voltage in your College?
2   What is the frequency of the supply, in cycles per second?
3   What is the name of the insulation cover which prevents electric shock on the welding cables?
4   What is the material called from which the earth clamp is made?
5   For welding purposes the voltage is comparatively low/high and the amperage is comparatively low/high.
6   When an arc is formed across a gap in an electric circuit, what state is the circuit said to be in?
7   It is important that the power leads must be well maintained, with sound efficient joints and prevented from being crushed, cut, burnt, stretched excessively and mishandled. State why you think this is important.

Fig. 14.2

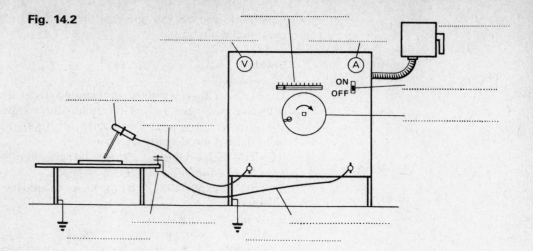

8 Draw the symbols for negative polarity and positive polarity.

9 Name the make and state the O.C.V. for an a.c. welding set in your College.

10 Name the make and state the O.C.V. for a d.c. welding set in your College.

11 What can you say about the current and the circuit when the arc voltage is measured?

12 Is a low voltage safety device fitted when welding under normal conditions or under hazardous conditions?

13 What kind of current do the following supply:

a) A generator

b) A transformer

c) A rectifier.

14 Poor contacts in the electrical circuit may cause

a) overheating

b) explosions

c) too much current flow

d) flashback.

15 Factors Affecting Metal Arc Welding

Variations in Current Values

If the welding current is too high, a flat wide bead with coarse ripples results from the increased arc force. The arc force also

Table 15.1 *Current Selection*

Diameter (mm)	Length (mm)	Current (amps)
8	450	300–500
6.3	450	200–370
6	450	190–310
5	450	150–250
4	450	120–180
4	350	120–190
3.25	450	80–125
3.25	350	80–130
2.5	350	60–95
2	300	50–80

produces a deep penetration pattern accompanied by an excessive amount of spatter. Though striking the arc is made easier, the crater is deep with blowholes and sometimes has cracks in the centre. Porosity (gas entrapped in the weld) is sometimes the result of too high a current. Overheating, particularly on thin section material being welded, and also of the electrode, results in unsatisfactory welded joints.

If the welding current is too low, the arc is difficult to control and often the electrode end fuses to the plate. This causes a short circuit and the electrode becomes red hot due to resistance heating unless it is broken off. Other features of welding with too low a current are: the weld bead tends to be high and globular and of irregular width with slag trapped in crevices and difficult to remove. The penetration is shallow in the centre of the bead whilst the toes of the weld are often just adhered to the plate.

Welding Investigation 6
Current Values and Electrode Size

AIM Selection of correct current values for electrode size.

THEORY For every electrode there is an optimum current value to suit the material and job being welded. When this value is arrived at, mainly by trial and error and experience, it is found that it is possible to deposit a weld bead free from the defects which would be caused by too low a current value or too high a current value. The range of current values should always be shown on the electrode box (Table 15.1). The correct welding current can be observed by using a tong test ammeter. The readings indicated on welding plant are not always accurate.

EQUIPMENT *Safety:* gloves, mask, apron.
Equipment: chipping hammer, wire brush, tongs, saw, macro equipment.
1 plate 150 mm × 100 mm × 10 mm.
General purpose 8 s.w.g. electrode (4 mm) Class 2.
Current: 145–190 amperes or from the electrode box (Table 15.1).
Preparation: Surface cleaned of rust, scale, grease, etc.

PROCEDURE
1) Mark three lines equidistant and lengthways with white chalk on the plate.
2) Set the lowest current value and deposit a bead of weld on the left-hand chalk line. Remember to keep a constant arc length. The length of the arc should be about the diameter of the core wire and angle of slope 65°–75° with angle of tilt 90°.

3) Allow to cool for a short time then deposit a bead on the centre chalk line using a current setting midway between the high and low currents.
4) Allow to cool and then finally deposit the last bead with the highest current.
5) When cool chip off the slag (care required) and saw cut across the three beads. One cut edge may then be prepared for macroscopic examination (page 58). Then draw and complete a table as indicated.

Current Value	Depth of Penetration	Relative Height and Width of Weld Bead	Spatter and Appearance of Ripples	Remarks
1A	———	———		

NOTES

1 It is possible that none of these current settings is the optimum one and further settings can be tried.

2 Variations in supply voltage can affect the current value and numerical values on the plant may not be the true amperage. A tong test ammeter can verify the amperage.

Variations in Voltage and Arc Length

As mentioned previously there is a striking voltage which initiates the arc, which once selected is fairly constant. The *arc voltage*, which is less than half the striking voltage, is *not* constant and fluctuates with the arc length during welding.

It should be noted that

1 When the arc length increases, the *arc voltage rises* and the *current decreases;*

2 When the arc length decreases, the *arc voltage decreases* and the *current rises;*

3 The total power output remains unaffected, i.e. volts × amps = watts (power).

Using General Purpose Electrodes

If the arc length is too long, the current is reduced and hence also the arc force. This results in a lack of penetration, increased spatter, a coarser weld bead, a brighter arc, loss of heat through dissipation, loss of some of the gaseous shield, and slag which results in an inferior weld deposit.

There are certain types of electrodes (touch type) with which it is impossible to maintain too short an arc under normal conditions. The general purpose electrode, if used with too

Table 15.2. (Procedure as for Investigation 6)

A) EXAMPLE OF CURRENT VALUES ON WELD BEAD

Current Value	Depth of Penetration from Macro	Relative Height of Weld Bead and Width	Spatter and Appearance of Ripples	Remarks
Optimum amps				Fine spatter, ripples smooth, round crater, smooth arc.

B) EXAMPLE OF A TOO HIGH CURRENT VALUE

Current Value	Depth of Penetration from Macro	Relative Height of Weld Bead and Width	Spatter and Appearance of Ripples	Remarks
Maximum amps				Coarse ripples, flat bead, excessive penetration, hollow porous crater, coarse excessive spatter, loud crackling arc.

C) EXAMPLE OF A TOO LOW CURRENT VALUE

Current Value	Depth of Penetration from Macro	Relative Height of Weld Bead and Width	Spatter and Appearance of Ripples	Remarks
Minimum amps				Little penetration, uneven width and height of bead. Irregular ripples, slag traps. Electrode tends to stick.

short an arc length, produces a weld bead which has heaped-up layers and often holes in the surface of the weld. These holes indicate that the weld metal has solidified at the instant the gas was escaping from the weld. An irregular and rounded bead, with slag flooding the arc in addition to the electrode "sticking", are symptoms of the arc length being too short.

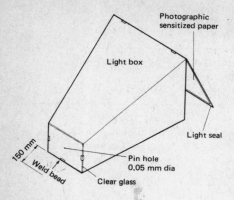

Photographic
sensitized paper

Light box

Light seal

150 mm

Weld bead

Pin hole
0.05 mm dia

Clear glass

Fig. 15.1 Light box

Welding Investigation 7
Variations in Voltage and Arc Length

EQUIPMENT As for Investigation 6, plus light box and sensitized paper. (Copyrapid negative medium C.R.M. 203× 330 mm. Developer AGFA. CR 160D 1.5 litre.)

PROCEDURE
Mark four equidistant chalk lines lengthways on the plate and using an automatic traverse, or handwelding using a voltmeter and ammeter, deposit four weld beads and draw and complete a table as shown.

Now complete a weld run with the light box as shown in Fig. 15.1. Develop and print, and compare print with weld.

Arc Length	Appearance of Bead Shape	Cross-section Macro	Arc Volts	180° Bend	Remarks
Too short					
Correct					
Too long					
Variation short to long					Use light box

Variations in Speed of Travel

It is important that the speed of travel is correct as this dictates the length of weld deposited per electrode, and from this other factors develop. For instance, excess heat concentration (slow travel) leads to wide flat deposits and to metal collapse on root runs. The slag often tends to run in front of the arc and make deposition difficult to control, especially when welding in the flat position.

When the speed of travel is too fast, the result is lack of penetration and fusion with a weld bead which is narrow and globular in section and tends to trap the slag in the toe of the weld. The ripples on the weld bead will show whether the speed was fast (elongated ripples) or slow (coarse and flat) similar to the major and minor axes of an ellipse.

Welding Investigation 8
Variations in Speed of Travel

EQUIPMENT As for Investigation 6, plus four plates, 150 mm × 100 mm × 6 mm.

PROCEDURE

1) Notch 70 mm of the electrode for each bead.
2) Mark four equidistant chalk lines on the length and deposit weld beads, keeping a constant arc length and allowing each bead to cool to room temperature.
3) Section and etch each plate across the bead and draw and complete the Table.

Speed of Deposition	Plan View of Appearance	Cross-section Macro	Length of Deposit (mm)	Remarks: Spatter and Type of Defect, etc.
A) Excessive				
B) Too slow				
C) Correct Example:			50 mm	Little spatter, fine. Slag removal very easy. No undercut.
D) Varying slow to fast				

Note: A 450 mm electrode should deposit approximately 225 mm of weld with 40 mm stub waste.

Variation in Angles of Electrode (Slope and Tilt)

For welding in the flat position it is important that the angle of tilt is 90° to ensure that the weld bead is deposited symmetrically (even about its centre line). There is also a tendency for cold laps to form on the one side and under-cutting to take place on the other side when the electrode is tilted too far to one side.

The electrode slope, which varies between 65° and 75°, affects the depth of penetration and also the height and shape of the weld bead. The steeper the angle the flatter the bead and the greater the tendency to penetrate, especially on single vee root runs. Slag also tends to flood the weld. On the other hand, a very shallow angle results in the arc pushing up the molten weld metal into a higher bead with less penetration, and danger of slag traps.

Welding Investigation 9
Variations in Angles of Electrode

PROCEDURE as for Investigation 8, with speeds as in Table on page 44. Complete Table as on page 74.

73

| Angles | | Plan View of | Cross-section | Remarks on Surface |
Slope	Tilt	Appearance	Macro	Imperfections
70°	70°			
50°	90°			
90°	90°			
70°	90°			

16   Electrodes and their Effect on Welds

Covered Electrodes for the Manual Metal-Arc Welding of Carbon and Carbon Manganese Steels (BS 639: 1976)

This British Standard is intended to show whether the Manufacturer has met the standards laid down for a particular job. It replaces the previous system of coding in BS 1719: 1963.

The first part of the coding is compulsory (up to the covering-type letter). The rest can be added at the manufacturer's discretion.

Examples: complete code E5133B16020(H)
             compulsory part E5133B.

The full explanation of the code is given in the diagram. Pages 74 to 78 are reproduced with kind permission of the British Standards Institution.

Characteristics of Electrode Coverings

A   Electrodes of the acid type have a medium or thick covering and produce an iron oxide/manganese oxide/silica slag, the metallurgical character of which is acid. The covering contains, besides oxides of iron and/or manganese, a fairly high percentage of ferro-manganese and/or other deoxidisers. The slag solidifies in a characteristic honeycomb structure and is easily detached.

This type of electrode usually has a high fusion rate and may be used with high current intensities. Penetration can be good, particularly if the covering is thick. These electrodes are most suitable for welding in the flat position but can also be used in other positions. Either direct or alternating current can be employed. With this type of electrode the weldability of the parent metal has to be good, as otherwise solidification cracking may occur. Susceptibility to solidification cracking is more

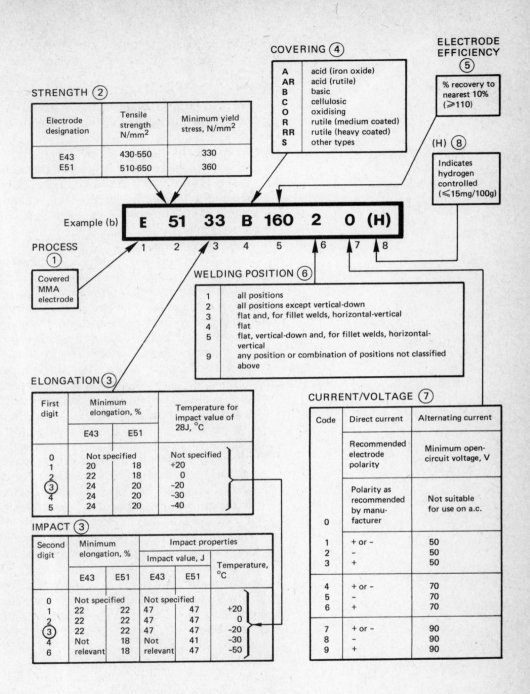

particularly marked in horizontal-vertical or vertical fillet welds when the actual carbon content exceeds approximately 0.24%, killed steel being more susceptible than rimmed steel, and when the sulphur content exceeds 0.05% in killed steels and 0.06% in rimmed steels.

AR Electrodes of the acid-rutile type usually have a thick covering, producing a slag very similar to the slag described under type A. Usually this slag is somewhat more fluid. The properties of a type AR electrode are in general very similar in all aspects to type A, the difference being that the covering contains titanium oxide, the amount normally being not higher than 35%.

Between the two types A and AR, several mixtures are possible, but when the amount of titanium oxide is larger than the total amount of iron and/or manganese oxide, the covering is considered to be of the AR type. If, instead of a mixture of iron oxide and titanium oxide, the mineral ilmenite is used, the same rule is applied.

B Electrodes of the basic type usually have a thick covering containing considerable quantities of calcium or other basic carbonates and fluorspar so that metallurgically they are basic in character. There is a medium quantity of dense slag, which often has a brown to dark-brown colour and a glossy appearance. It is easily detached, and as it rises to the surface of the weld very quickly, slag inclusions are not likely to occur. This type of electrode gives an arc of average penetration, and is generally suitable for welding in all positions. This type of electrode is often used on direct current, electrode positive, but there are electrodes that can be used on alternating current.

As the deposited metal is highly resistant to solidification and cold cracking, these electrodes are particularly suitable for welding heavy sections and very rigid carbon steel structures. They are also recommended for welding medium-tensile steels and steels, the carbon and sulphur content of which are higher than those of carbon steel of good weldable quality.

The coverings of basic electrodes have to be very dry; consequently these electrodes need to be stored in a very dry place or, if they have already absorbed moisture, to be dried before use, according to the recommendation of the manufacturer. This ensures that the deposited metal will have a low hydrogen content, and there is less risk of underbead cracking when welding steels likely to show a marked hardening in the heat-affected zone.

C Electrodes of the cellulosic type have a covering which contains a large quantity of combustible organic substances, so that the decomposition of the latter in the arc produces a voluminous gas shield. The amount of slag produced is small and the slag is easily detached.

This type of electrode is characterized by a highly penetrating arc and fairly high fusion rate. Spatter losses are fairly large and the weld surface is somewhat coarse, with unevenly spaced ripples. These electrodes are usually suitable for welding in all positions.

O   Electrodes of the oxidising type have a thick covering composed mainly of iron oxides with or without manganese oxides. The covering gives an oxidising slag, so that the deposited metal contains only small amounts of carbon and manganese. The slag is heavy, compact and often self-detaching. This type of electrode gives poor penetration and a fluid molten pool, and is particularly suitable when only a small weld is required. Usually its use is restricted to welding in the horizontal-vertical fillet weld and flat fillet weld positions.

These electrodes are used mainly for welding steels when the appearance of the weld is more important than the mechanical strength of the joint.

R and **RR**   Rutile type electrodes have a covering containing a large quantity of rutile or components derived from titanium oxide. Usually this amounts to 50% by mass (not taking into account cellulosic material).

The R type can be distinguished from the AR type by its heavier slag. Sometimes this difference is not readily discernible, especially with electrodes having a medium covering, but the amount of rutile in the covering is well above 45%.

Because of the difference in application and also mechanical properties, a subdivision is made according to the thickness of the covering:

R   The covering is of medium thickness*. Small amounts of cellulosic material, up to the maximum of 15%, may be present in the covering. These electrodes are particularly suitable for welding in the vertical and overhead positions.

RR   The covering is of heavy thickness*. Small amounts of cellulosic material, up to a maximum of 5%, are sometimes present in the covering. The slag is heavy, compact and usually self-detaching. The appearance of the weld resembles that of an O type electrode.

Although the susceptibility to solidification cracking because of dilution by the parent metal is not so high as that of the acid type, care has to be taken in view of the fact that usually a weld is made with a much smaller throat thickness than with acid electrodes. The maximum current to be used is lower than that of an AR type, because of a lower melting rate.

EXAMPLES
The following examples illustrate the way in which the coding is expressed and the use of the complete classification or only the compulsory part.

---

* The limit between coverings of medium and heavy thickness corresponds approximately to a ratio of 1:5 between the outer diameter of the covering and the core wire diameter.

Example (a). Covered electrode for manual metal-arc welding having a rutile covering of medium thickness and depositing weld metal with the following minimum mechanical properties.

Tensile strength: 500 N/mm²

Elongation: 23%

Impact strength: 71 J at +20°C, 37 J at 0°C, 20 J at −20°C.

It may be used for welding in all positions. It welds satisfactorily on alternating current with a minimum open-circuit voltage of 50 V and on direct current with positive polarity.

The complete classification for the electrode would therefore be E 43 21 R 1 3 and the compulsory part would be 3 43 21R.

Covered electrode for manual metal-arc welding ——————————————

Tensile strength ——————————————

Elongation and impact strength ——————————————

Covering ——————————————

Welding positions ——————————————

Current and voltage ——————————————

Example (b). An electrode for manual metal-arc welding having a basic covering, with a high efficiency and depositing weld metal containing 8 ml of diffusible hydrogen per 100 g of deposited weld metal with the following minimum mechanical properties.

Yield stress: 380 N/mm²

Tensile strength: 560 N/mm²

Elongation: 22% } Also a minimum elongation of 20%

Impact strength: 47 J at −20°C } with an impact value of 28 J at −20°C

Nominal efficiency: 158%

It may be used for welding in all positions except vertical-down, direct current only.

The complete classification for the electrode would therefore be E 51 33 B 160 2 0 (H) and the compulsory part would be E 51 33B.

Covered electrode for manual metal-arc welding ——————————————

Tensile strength and yield stress ——————————————

Elongation and impact strength ——————————————

Covering ——————————————

Efficiency ——————————————

Welding positions ——————————————

Current and voltage ——————————————

Hydrogen controlled ——————————————

EXERCISE

Indicate the complete classification for a manual metal arc electrode having a rutile heavy coating which is used for welding flat and horizontal vertical fillet welds. It welds satisfactorily on + or − d.c., and a.c. with a minimum O.C.V. of 70 V. Yield stress 350 N/mm². Tensile strength 450 N/mm². Elongation 21%. Impact strength 47 J at +20°C.

Welding Investigation 10
Some Types of Electrode Coating

AIM Investigation of the characteristics of electrode coverings.

PROCEDURE

Mark out a fresh plate and, using various classes of electrode, deposit weld beads using the correct procedure (see pages 68 to 73) and then draw and complete the table.

Electrode covering 4 mm	Fumes	Position and Current	Type of Slag Underside	Ease of Slag Removal	Amount of Spatter	Penetration	Appearance of Bead	Undercut
A **AR** Example	White deposit on plate	160 amp Flat		Very good	Moderate	Good	Good	Small traces
B								
C								
O								
R								
RR								
S								

Welding Investigation 11
Size of Material (Including Thickness Limits)

THEORY The thermal effects using the same size electrode and current differ widely with different sizes and thicknesses of material.

Briefly, with thin plate, overheating and a wide flat bead and excessive penetration result with "burning" through unless the rate of travel is rapidly increased.

With thick plates the heat supplied now is not enough as it is dissipated through the thick metal. The weld bead is now rounded with root penetration and may be slightly harder. Cracking occurs in some cases.

AIM To investigate the effect of plate thickness on the weld.

MATERIAL   Three plates (150 mm × 100 mm × 3 mm, 6 mm, 12 mm).
4 mm electrode, general purpose, class 2—current 160 amps. Plates clean and free from rust and dirt.

PROCEDURE
Deposit a bead on the centre of each plate, clean, section and macro, and draw and complete the Table.

Plate	Appearance of Bead	Sections Macro Start	Finish	Remarks
Thin 3 mm				
Medium 6 mm				
Thick 12 mm				
Thick 12 mm 195 A				

Welding Investigation 12
The Effects on Welds of Using Defective Electrode Coatings

THEORY   It is important that electrodes are stored in a warm dry place, approximately 10 to 15°C with atmospheric humidity not above 40%, and used in rotation, i.e. new electrodes are not piled on top of old stock. Sometimes very old electrodes have a coating of white fur caused by water glass in the electrode. Electrodes should be carefully stacked and not dropped which would cause cracking and chipping of the coating. Poor quality welds usually result if chipped, damp or defective electrodes are used. If dry electrodes are shaken in the hands they produce a hard metallic sound, but damp ones have a hollow sound.

AIM   To investigate the effect of using defective electrodes.

EQUIPMENT   As for Investigation 6 (page 69).

PROCEDURE
Take an electrode with the coating chipped off in parts so that approximately 15 mm of bare wire shows every 75 mm. Ensure that the first 75 mm is soaked in water, the next 75 mm is smeared in oil, and the next 75 mm is bare wire.

Deposit a weld bead, then section along the *length* and macro etch (see below) and then draw and complete the table. Deposit a bead now with a very old electrode and section along, then macro.

START　WATER　BARE　　　OIL　　BARE　　　AS
　　　　SOAK　ELECTRODE SMEAR ELECTRODE RECEIVED

PROFILE OF
BEAD MAXIMUM
PENETRATION

LONGITUDINAL SECTION (MACRO)

Condition	†Arc	*Amount of Spatter	*Blow Holes	*Under-cut	*Penetration	Shape of Bead	*Overlap	*Porosity
Water soak								
Bare electrode								
Oil smear								
As received								
Very old								

* *Note:* Mark—none (N), little, (L), fair (F), excessive (Ex).
† *Note:* Mark—fierce (F), Normal (NI), erratic (Er).

Welding Investigation 13
Effect of Electrode Polarity

THEORY　When a direct current carbon or tungsten arc is struck, the approximate proportion of the arc heat is one-third at the negative pole and two-thirds at the positive pole. This may be verified by sectioning the plate and observing the depth of penetration for each pole. With an a.c. carbon arc the arc heat is 50% at either pole.

The above is not always true when using coated electrodes as the burn-off rate (speed of melting) and also the depth of penetration vary, depending on the electrode coating (i.e. class of electrode).

The efficiency and the position of the welding return can also affect the burn-off rate and depth of penetration, due to erratic arc conditions.

AIM　To investigate the effects of electrode polarity.

EQUIPMENT　As for Investigation 6 (page 69).

MATERIAL　Quantity of class B general purpose electrodes 4 off at 4 mm dia. at 160 amps. Plate 150 mm × 100 mm × 10 mm (two off).

PROCEDURE

Mark off 150 mm from the end of each electrode and saw cut around the flux coating. Check that 160 amps is correct (tong test ammeter or ammeter on welding set).

Deposit the 150 mm of each electrode on four chalk lines drawn equidistant lengthways. Allow each bead to cool before depositing the next. Remove slag (care).

Measure each bead, section and macro etch. Then draw and complete the Table.

Type of Current and Pole	Marked 15 cm Electrode Deposit Length	Time (sec)	Cross-section Macro	Bead Shape	Remarks
(1) D.C. −ve					
(2) D.C. +ve					
(3) A.C. Lead					
(4) A.C. Return					

Note: For a.c. plant, change current return cable to lead socket.

17   Control of Distortion (Welding)

The majority of steel plate and sections have "locked in" stresses caused by hot rolling followed by uneven cooling at the steel mills. There is an increase in stress if any cold forming operations are subsequently carried out, such as rolling, pressing and shearing, etc. Thermal cutting also leaves stresses at the cut edge. These stresses, which are present before any welding is carried out, are usually referred to as *residual stresses* and, coupled with other factors mentioned below, cause displacement of members out of alignment.

When a thick plate material is being welded, the heated metal expands and tries to force the plates apart, but is restrained to a large extent by the cold surrounding material. As the weld progresses, the weld metal begins to contract and exerts a shrinkage force which pulls the plates together. This force combined with the high temperature of the adjacent metal causes the material to YIELD and deform PLASTICALLY in the area affected by the heat. This causes internal stresses to be set up, also known as residual stress. In certain cases the resulting stresses will also cause cracking either during welding or in subsequent service.

When welding thin plate, the distortion is usually seen as buckles or bowing of the plate.

There are certain well tried methods of controlling distortion both before and during welding and the more common methods now follow.

Before Welding

1 *a*)   Use correct welding procedure sheets (BS 499; 5135, 5500) and ensure a correct edge preparation and fit-up. Use double-vee or double-U where flatness is important for thick plates. (Double-U has the least volume of weld metal.) A 2 mm root gap is reasonable to ensure penetration.

b)   Ensure that plates are not misaligned and choose a welding process which produces the least distortion—for example, electroslag for thick slab, tags-mags or spot welding rather than oxy-acetylene for thin sheet.

2 *a*)   Make shrinkage forces (see below) work to achieve correct alignment by the use of pre-setting.

PRE-SETTING   Locate parts in such a position that they pull into line due to the contracting weld metal. Examples are shown in Fig. 17.1. The shrinkage direction is shown by the arrows. This method is employed on sub-assemblies as the parts have complete freedom to move. Restraint methods (see page 84) are usually used for final assembly.

To minimise stress, the direction of welding should be away from the point of restraint to the point of maximum freedom. A correct tacking formula should be used for the particular metal being welded if tacks are used.

b)   PRE-CAMBERING   The plates to be welded are kinked in a press, or rolled slightly, or cambered using clamps to

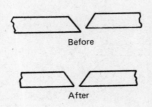

Before

After

Fig. 17.1

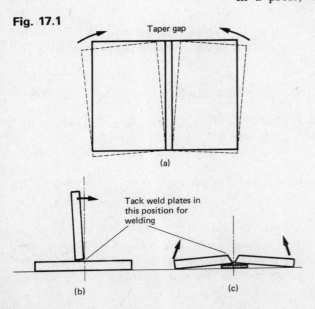

Taper gap

(a)

Tack weld plates in this position for welding

(b)

(c)

Fig. 17.2

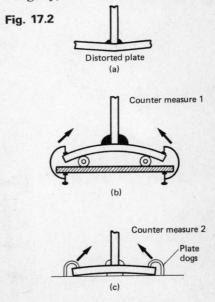

Distorted plate
(a)

Counter measure 1

(b)

Counter measure 2

Plate dogs

(c)

counter the shrinkage. Fig. 17.2 shows examples of distortion and countermeasures.

3 BACK-TO-BACK ASSEMBLIES (to avoid bowing) Identical or similar components are used to restrain and balance the weld shrinkage of each other, when fastened back to back. Typical examples are given in Fig. 17.3a, b and a balanced weld sequence is shown at c.

Fig. 17.3

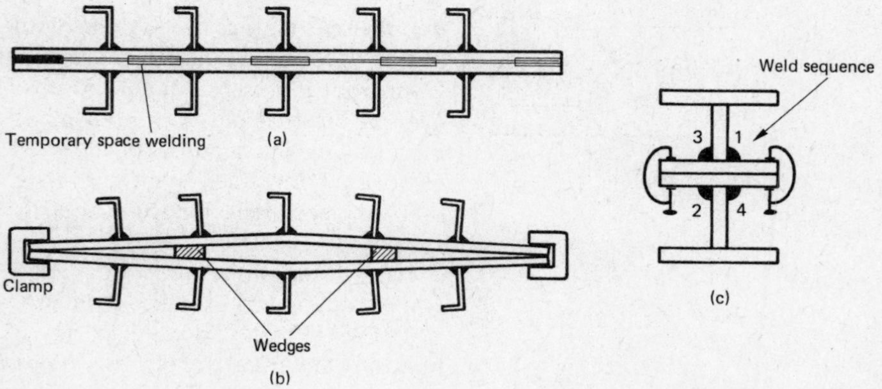

Temporary space welding   (a)

Weld sequence

Clamp

3 1
2 4

Wedges

(b)

(c)

Fig. 17.4

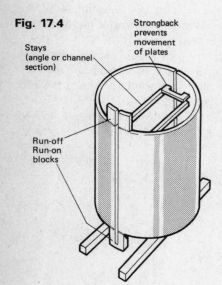

Strongback prevents movement of plates

Stays (angle or channel section)

Run-off Run-on blocks

4 RESTRAINING METHODS Jigs and fixtures may be used to restrain or control movements of the components during welding. If close control over accuracy is required, then restraint of the workpiece may be achieved by using wedges, strongbacks, chains, clamps and stays, which remain in position during cooling. As these methods prevent most of the movement, there is progressive build-up of stress in the fabrication. Where necessary, stress relief is carried out with the stays, etc. in position (Fig. 17.4).

Methods of Countering Weld Shrinkage (before and during welding)

Longitudinal and transverse shrinkage are shown in Fig. 17.5.

1 BEFORE WELDING Typical workshop allowances for weld metal shrinkage are:

Longitudinal
Fillet welds: 0.8 mm per 3 m of weld
Butt welds: 3.0 mm per 3 m of weld

Transverse
Fillet welds: 0.8 mm per weld where leg length does not exceed three quarters of plate thickness.
Butt welds: 1.6 mm per weld for 60° vee joints.

Fig. 17.5 Longitudinal and transverse shrinkage

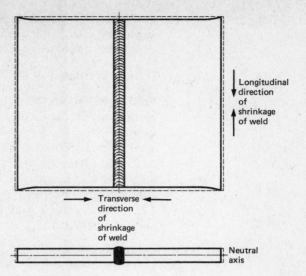

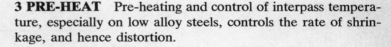

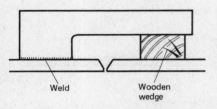

Weld   Wooden wedge

Fig. 17.6 Arrangement of clamp to prevent angular distortion while permitting transverse shrinkage

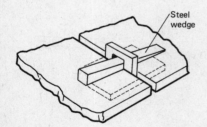

Steel wedge

Fig. 17.7

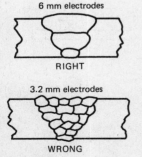

6 mm electrodes

RIGHT

3.2 mm electrodes

WRONG

Fig. 17.8

2 CLAMPS AND WEDGES Fig. 17.6 illustrates a method of controlled shrinkage. Use may be made of spacer wedges (Fig. 17.7), clamps and cleats to hold plates in alignment. They are removed as the welding progresses or after completion.

3 PRE-HEAT Pre-heating and control of interpass temperature, especially on low alloy steels, controls the rate of shrinkage, and hence distortion.

4 CHILLS Chills, although used extensively when welding thin plate, are only occasionally used in thick plate welding. They are used to extract heat quite rapidly from a specific place on the weldment, to prevent "burn through" or to prevent heat spread at critical places. Chills may be in the form of copper, water-cooled backing pieces, soft aluminium blocks or machined cast iron supports.

5 WEAVING Excessive weaving and deposition of excess weld metal in any part of the fabrication at one time should be avoided. It is good practice to limit the weave to no more than 3 times the electrode diameter.

6 EXCESS WELD METAL Avoid over-welding. This means that weld metal should be kept to a minimum consistent with requirements. It is a costly mistake to deposit excess-weld metal, as this increases distortion but not strength. Use the minimum number of runs unless otherwise stated. Fig. 17.8 gives an example.

7 WELDING SPEED The optimum speed of travel should be used because too slow a speed tends to increase distortion.

8 PEENING Careful peening between runs promotes a stretching effect which counteracts the shrinkage. Used especially when welding cast iron.

During Welding

WELD SEQUENCE The order in which welds are to be carried out should be clearly marked on the fabrication before welding commences. A *balanced welding sequence* should be used so that successive weld runs pull against each other from each side of the neutral axis to retain alignment. String lines and plumb bob will show relative movement. Examples are shown in Fig. 17.9 of two methods of sequence welding. Fig. 17.10 shows how a balanced sequence may be used to avoid angular distortion.

Fig. 17.9

General direction of welding

Direction of welding each electrode

Above back-step welding

| 1 | 2 | 3 | 4 | 5 | 6 | 7 | 8 | 9 |

| 1 | 6 | 2 | 7 | 3 | 8 | 4 | 9 | 5 |

Below skip welding or planned wandering

Fig. 17.10

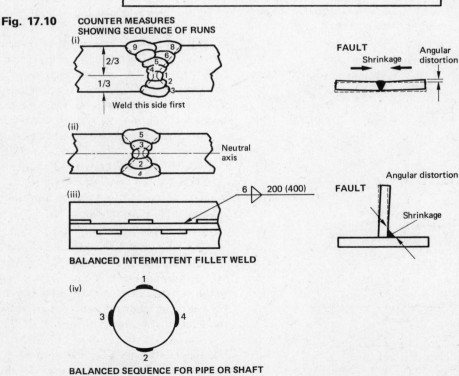

COUNTER MEASURES SHOWING SEQUENCE OF RUNS

(i)

2/3

1/3

Weld this side first

FAULT Shrinkage Angular distortion

(ii)

Neutral axis

(iii)

6 ▷ 200 (400)

BALANCED INTERMITTENT FILLET WELD

FAULT Angular distortion Shrinkage

(iv)

BALANCED SEQUENCE FOR PIPE OR SHAFT

After Welding

REMOVING DISTORTION It must be remembered that it costs money to remove distortion after welding. Care must be taken not to introduce too much additional stress especially when pulling and twisting.

1 MECHANICAL METHODS See Fig. 17.11.

A certain amount of distortion may be rectified by mechanical means such as pressing, jacking, hammering, bending rolls or crane lift, taking care not to impose undue stress on the welds which could cause cracking. Local heating may be necessary to allow easier movement.

Fig. 17.11 Mechanical methods of removing distortion

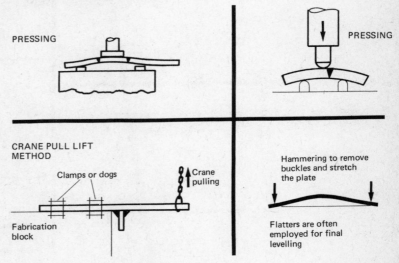

PRESSING

PRESSING

CRANE PULL LIFT METHOD

Clamps or dogs

Crane pulling

Fabrication block

Hammering to remove buckles and stretch the plate

Flatters are often employed for final levelling

2 THERMAL METHODS

A component which is buckled may be straightened without mechanical means by heating in local zones on the convex side, starting at the centre of the bulge and working progressively outwards. Wedge-shaped zones should be heated to a dull red colour and unrestrained contraction allowed to take place. Examples are given in Fig. 17.12. The following points should be observed:

1 The proposed heating zone should be marked out and heating commenced at the base, working towards the apex. A dull red heat should not be exceeded, approx. 700°C for steel.

2 The wide part or base of the wedge should be at the outer edge and should be of a width of approximately one third of the length of the wedge. The apex or point of the wedge should reach the neutral axis of the assembly.

3 An intense flame should be used to heat the zone rapidly before dispersal of heat into the surrounding areas occurs. Oxy-acetylene preferred.

Fig. 17.12
Thermal methods of removing distortion

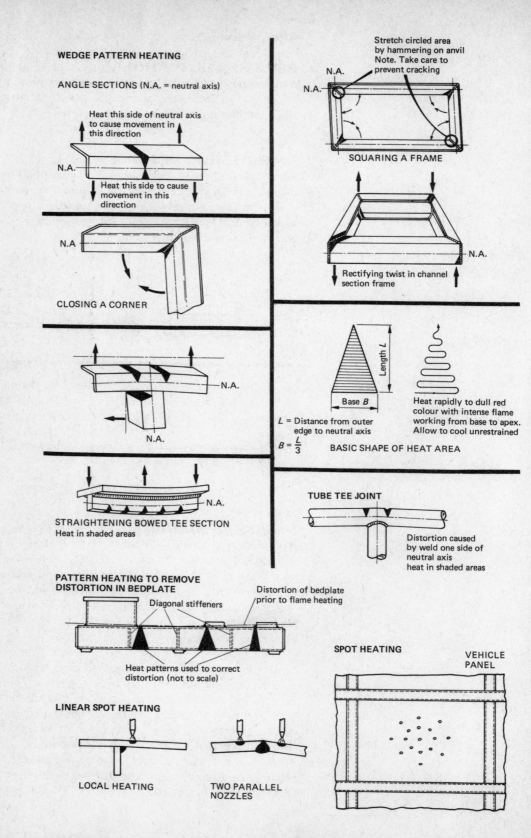

WEDGE PATTERN HEATING

ANGLE SECTIONS (N.A. = neutral axis)

Heat this side of neutral axis to cause movement in this direction

N.A.

Heat this side to cause movement in this direction

N.A

CLOSING A CORNER

N.A.

N.A.

STRAIGHTENING BOWED TEE SECTION
Heat in shaded areas

N.A.

Stretch circled area by hammering on anvil
Note. Take care to prevent cracking

N.A.

N.A.

SQUARING A FRAME

N.A.

Rectifying twist in channel section frame

Length *L*

Base *B*

L = Distance from outer edge to neutral axis

$B = \dfrac{L}{3}$

Heat rapidly to dull red colour with intense flame working from base to apex. Allow to cool unrestrained

BASIC SHAPE OF HEAT AREA

TUBE TEE JOINT

Distortion caused by weld one side of neutral axis heat in shaded areas

PATTERN HEATING TO REMOVE DISTORTION IN BEDPLATE

Diagonal stiffeners

Distortion of bedplate prior to flame heating

Heat patterns used to correct distortion (not to scale)

LINEAR SPOT HEATING

LOCAL HEATING

TWO PARALLEL NOZZLES

SPOT HEATING

VEHICLE PANEL

4   Where an increased effect is required, the surrounding areas may be cooled, i.e. water jet on the reverse side of the heated portion.

5   This method must not be used on hardenable steels.

Note that spot heating is similar to wedge heating; quenching with water is permissible on non-hardenable steels.

18   Tungsten Arc Gas Shielded Welding

This process is a method of welding whereby an electric arc is maintained between a virtually non-consumable tungsten electrode, in an atmosphere of pure argon, with or without small additions of other beneficial gases. The gas shield prevents weld metal contamination by the atmosphere. The surface of aluminium alloys is covered by a refractory high melting point film of oxide which must be removed before a satisfactory weld can be made. A filler wire may also be added at the leading edge of the molten pool to form the weld. It is one of the characteristics of an alternating current arc that it removes this tenacious oxide during the welding process. The illustrations in the table show the characteristics of a.c. and d.c. arcs (page 90).

Arc Starting

To initiate the arc for welding the two most common methods are:

High Frequency (H.F.) A series of high-voltage high-frequency sparks are superimposed on the main welding current so that, at the press of a switch, they pass from the tungsten to the work and so *ionise* the air gap (break down the resistance) and allow the welding current to create an arc. This avoids touching the plate with the tungsten and avoids contamination. The H.F. may be continuous for a.c., and for d.c. used only when the arc has been extinguished.

Surge Injection is another method of arc starting which also uses a high-voltage high-frequency spark.

Fig. 18.1 shows a water-cooled torch. Gas-cooled torches are used for currents up to 150 A. Fig. 18.2 shows the system.

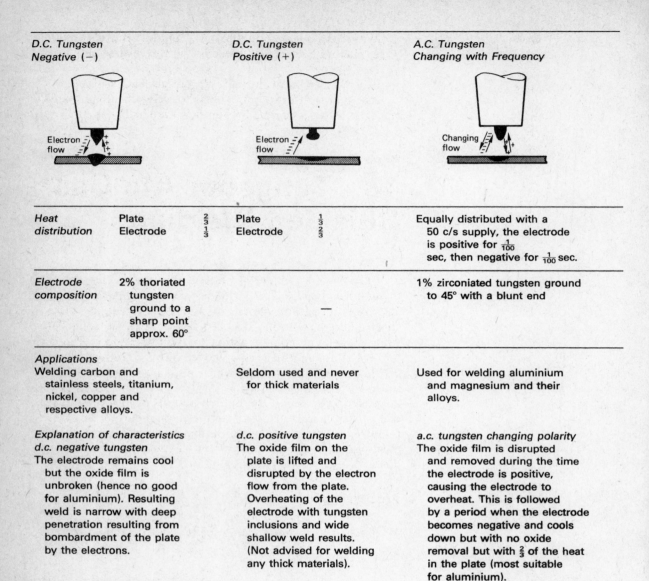

	D.C. Tungsten Negative (−)	D.C. Tungsten Positive (+)	A.C. Tungsten Changing with Frequency
Heat distribution	Plate $\frac{2}{3}$ Electrode $\frac{1}{3}$	Plate $\frac{1}{3}$ Electrode $\frac{2}{3}$	Equally distributed with a 50 c/s supply, the electrode is positive for $\frac{1}{100}$ sec, then negative for $\frac{1}{100}$ sec.
Electrode composition	2% thoriated tungsten ground to a sharp point approx. 60°	—	1% zirconiated tungsten ground to 45° with a blunt end
Applications	Welding carbon and stainless steels, titanium, nickel, copper and respective alloys.	Seldom used and never for thick materials	Used for welding aluminium and magnesium and their alloys.
Explanation of characteristics	*d.c. negative tungsten* The electrode remains cool but the oxide film is unbroken (hence no good for aluminium). Resulting weld is narrow with deep penetration resulting from bombardment of the plate by the electrons.	*d.c. positive tungsten* The oxide film on the plate is lifted and disrupted by the electron flow from the plate. Overheating of the electrode with tungsten inclusions and wide shallow weld results. (Not advised for welding any thick materials).	*a.c. tungsten changing polarity* The oxide film is disrupted and removed during the time the electrode is positive, causing the electrode to overheat. This is followed by a period when the electrode becomes negative and cools down but with no oxide removal but with $\frac{2}{3}$ of the heat in the plate (most suitable for aluminium).

Applications

The T.A.G.S. welding process is used where high-quality neat-looking welds are required, and is economical for thicknesses up to 6 mm. For thicknesses greater than 6 mm, M.A.G.S. welding is usually used or other metal arc welding processes. Root runs in pipe joints either with or without fusible inserts are put in using T.A.G.S. welding because the penetration can be controlled to give a smooth flush finish. After inspection the remainder of the joint is completed using quicker methods. Fig. 18.3 shows the electrode shapes used.

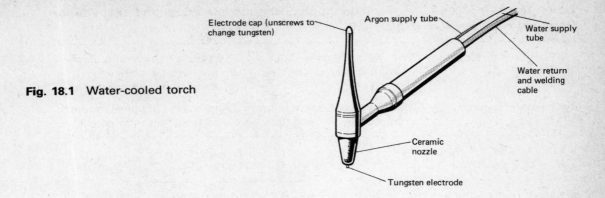

Electrode cap (unscrews to change tungsten)

Argon supply tube

Water supply tube

Water return and welding cable

Ceramic nozzle

Tungsten electrode

Fig. 18.1 Water-cooled torch

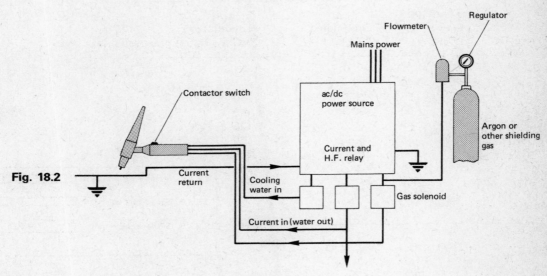

Fig. 18.2

Contactor switch

Mains power

Flowmeter

Regulator

ac/dc power source

Current and H.F. relay

Argon or other shielding gas

Current return

Cooling water in

Gas solenoid

Current in (water out)

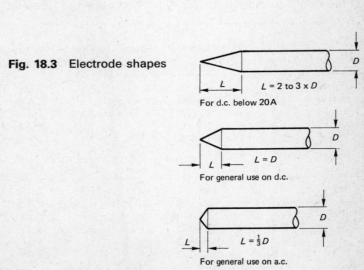

Fig. 18.3 Electrode shapes

$L = 2$ to $3 \times D$

D

For d.c. below 20A

$L = D$

D

For general use on d.c.

$L = \frac{1}{3}D$

D

For general use on a.c.

L

Table 18.1 Application of Two Common Electrode Materials and the Various Electrode Point Preparations

1–2% Thoriated tungsten	Mainly d.c. negative but possible on d.c. positive	Ferrous materials and some non-ferrous excluding aluminium and magnesium
Zirconiated tungsten	Mainly a.c. but possible on d.c. positive or negative	Aluminium, magnesium and alloys of these materials

Table 18.2 Current Ratings for Tungsten Arc Electrodes

	Maximum current carrying capacity in amps			B.S. recommendations for Class 1 welds
	Thoriated		Zirconiated	Zirconiated
mm	D.C.	A.C.	A.C.	A.C. (with suppressor)
0.8	45	30	—	—
1.2	70	40	40	—
1.6	145	55	55	50
2.4	240	90	90	80
3.2	380	140	150	120
4.0	440	195	210	160
4.8	500	250	275	200
5.6	—	275	320	250
6.4	—	320	370	320
7.9	—	410	—	—
9.5	—	500	—	—

S.E.C.B. Square edge close butt
S.E.O.B. Square edge open butt — 1.6 mm
S.V.C.B. Single vee close butt — 1.6 mm

Table 18.3 Selection of Welding Variables for Specific Joint Preparations

Material	Thickness (mm)	Current (A)	Type of Power	Argon Flow (1./min)	Electrode Dia. (mm)	Nozzle Bore (mm)	No. of Passes	Prep. Type (see above)
Stainless, heat and corrosion resistant	1.2	40– 60	d.c. or a.c.	2.5	1.6 or 2.4	6	1	S.E.C.B.
	1.6	60– 80	d.c. or a.c.	3	1.6 or 2.4	10	1	S.E.C.B.
	2.0	80– 90	d.c. or a.c.	3.3	2.4	10	1	S.E.C.B.
	2.6	90–110	d.c.	3.5	2.4 or 3.2	10	1	S.E.C.B.
	3.2	110–130	d.c.	4	2.4 or 3.2	10 or 12	1	S.E.O.B. S.V.C.B. 80°
	4.8	130–170	d.c.	4	3.2	10 or 12	1	S.V.C.B. 80°
Aluminium and alloys	1.2	60– 70	a.c.	3.3	2.4	10	1	S.E.C.B.
	1.6	70– 90	a.c.	4.7	2.4	10	1	S.E.C.B.
	2.0	90–110	a.c.	4.7	3.2	12	1	S.E.C.B.
	2.6	110–130	a.c.	5	3.2	12	1	S.E.C.B.
	3.2	130–150	a.c.	6	3.2	12	1	S.E.C.B.
	4.8	150–200	a.c.	7	4.8	12	1	S.E.C.B.

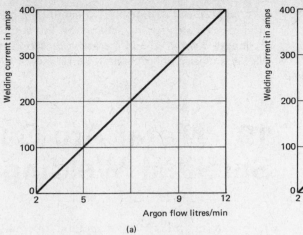

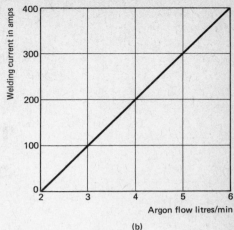

(a)

(b)

Fig. 18.4 (*a*) Aluminium and Alloys and Magnesium Alloys. (manual welding)
B.S. 3019: Part 1 1958 incorporating amendments (P.D. 4574) and (P.D. 5875)
(*b*) Austenitic Stainless and Heat-resisting steels. (manual welding)
B.S. 3019: Part 2 1960 incorporating amendment (P.D. 4207)

Fig. 18.4 gives examples of general recommendations for T.A.G.S. welding of plate, butt welds. The rate of flow of argon necessary to obtain a clean weld will depend on several factors, such as parent material, shape and size of nozzle, type of joint and whether the work is to be done in the shop or on site. In general there is a relationship between argon flow and welding current.

Power Source

The power source used for d.c. welding of stainless steel is usually either a transformer rectifier or a motor generator set with the addition of an H.F. unit. For welding aluminium and magnesium alloys a transformer is used, again coupled with an H.F. unit. Conventional manual metal arc plant may be used with a separate H.F. unit, but it is better to use a custom-built plant complete with a suppressor unit, contactor control (for switching the current and H.F. on and off where necessary) and with gas and water flow controls.

The power source characteristics for T.A.G.S. welding are similar to manual metal arc welding, i.e. only a small variation in the welding current results from a fairly large increase in arc voltage, due to increasing the arc length. This arrangement allows a constant current despite fluctuations of the arc length due to the operator. Fig. 18.5 shows an example of a "drooping" characteristic as used in T.A.G.S. welding.

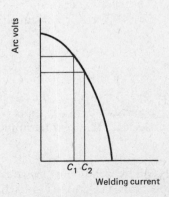

Fig. 18.5

It should be noted that a small change in arc voltage produces a much smaller change in arc current, i.e. when the arc length is reduced and the arc volts drop from V_1 to V_2 then the current rise is only small, i.e. C_1 to C_2.

19 Metal Arc Gas Shielded Welding

This welding process is a method of welding whereby an electric arc is maintained between a continuously fed consumable wire electrode. The protective gas shield, the wire and cooling water when necessary are fed through a flexible hose connected to the torch or gun at one end and the control unit at the other. This control unit usually houses electronic switches which stop and start the wire feed, the shielding gas flow and the cooling water, in addition to current and voltage control. A contactor switch usually on the gun causes the wire to feed through the copper contact tip in the end of the gun and so allow the arc to be struck.

It is important to note that the wire tip will only arc during the time the wire is feeding out, and *increasing* the wire speed causes an *increase in current*. The electrode is fed at a *constant speed* when selected at the control unit, but as stated above this speed may be varied to increase or decrease the current.

Power Source

Direct current using either a rectifier or generator is used in the M.A.G.S. welding system with the polarity of the electrode being positive. The power source characteristic is a "flat" power source as shown at Fig. 19.1 for a constant potential machine.

Modes of Metal Transfer

The mode of metal transfer from the tip of the electrode to plate may be influenced by current density, type of parent metal and electrode, gas shield, etc. The two basic modes are dip and spray transfer.

DIP TRANSFER This is where relatively *low arc voltages* and *currents* are utilised to obtain the smooth detachment of droplets from the electrode tip, utilising the electrode diameter

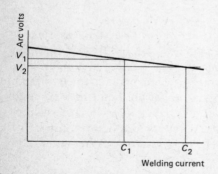

Fig. 19.1

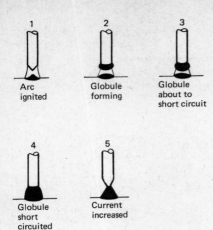

Fig. 19.2 Dip transfer

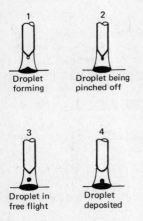

Fig. 19.3 Spray transfer (one droplet only is shown for clarity)

and arc voltage for a selected wire speed to give correct welding conditions. The size of the droplets increases as the current is reduced. "Stubbing", or welding of the tip to plate, occurs when current rise is too low. This mode of transfer is used for welding thin plate or sections and for positional work. The sequence of metal transfer is shown at Fig. 19.2.

SPRAY TRANSFER This mode of transfer is when the tip of the electrode is deposited in the form of a fine spray of molten metal droplets detaching smoothly at a high frequency (Fig. 19.3). True spray transfer occurs at relatively high arc voltages and low currents, but, due to the low melting point of aluminium, spray transfer can be obtained at relatively low current levels compared with steel. Fig. 19.4 shows a typical graph indicating spray and dip transfer ranges with the threshold range in between the two.

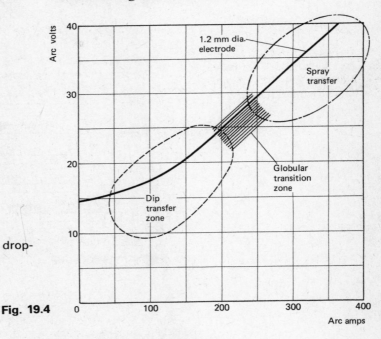

Fig. 19.4

Types of M.A.G.S. Torches

There are several types of torch but they may be divided into the gas-cooled and water-cooled types. The drive may be by electric motor with the wire spool on the hand-held gun, by air motor, or simply by a wire-feed push gun. A gas-cooled light-duty swan-neck torch is shown in Fig. 19.5.

Safety Precautions (T.A.G.S. and M.A.G.S.)

The safety precautions to be observed with these processes are similar for other metal arc processes with certain modifications.

Fig. 19.5 Light-duty swan-neck torch

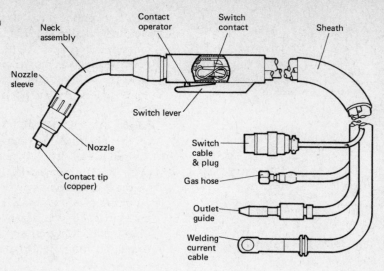

Table 19.1 General Guide to the Selection of Electrode Diameters for Low Carbon Steel Electrodes using Carbon Dioxide Shields

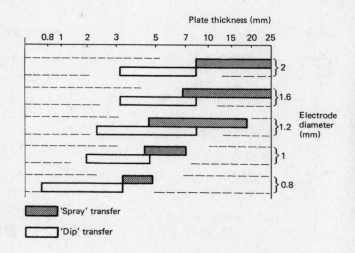

In confined spaces, gas shields if allowed to escape may displace oxygen and cause suffocation. Degreasing agents such as trichloroethylene and carbon tetrachloride decompose around the arc to form poisonous compounds. Local fume extraction should be used when employing very high current densities or flux core electrode wire, and filter breathing pads to prevent inhaling oxide dust. Correct grades of screen glass should be used as ultra violet light is greater when welding aluminium with an argon shield compared with other processes. Remember to chalk HOT on materials after welding, especially aluminium. Use light gloves when T.A.G.S. welding to avoid burning through radiation and H.F. burns between the fingers. Adequate protective clothing should always be worn.

20 Submerged Arc and Electro-slag Mechanised Welding

Submerged Arc Welding

This is a metal arc process in which an arc is struck between a copper-coated bare wire electrode and a rust-free workpiece under a blanket of granulated flux. This flux when cold is electrically non-conductive but when molten is highly conductive and allows very high welding currents to be used. The flux near the arc melts and provides protection for the weld pool from nitrogen and oxygen in the atmosphere and also concentrates the arc causing a deep penetrated weld of the highest quality. The flux floats to the surface and solidifies as a protective slag forming a smooth weld surface free from ripples. Surplus flux may be recovered, sifted and re-used.

Referring to Fig. 20.1, the welding feed wire is fed into a copper contact guide nozzle by two drive rolls which may also supply the current to the wire. The arc length is controlled automatically and the most common control method is the fully self-adjusting arc system employing a variable speed motor. The wire feed motor speed is controlled by the arc which acts as a variable resistance. The longer the arc the higher the arc voltage, so causing an increase in the speed of

Fig. 20.1 Submerged arc welding

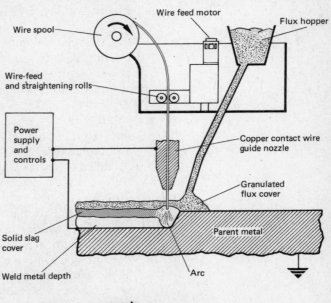

Welding direction

the motor and hence the wire speed, so correcting the arc length.

The other method is as for M.A.G.S. welding where the motor and wire speed remain constant. With a long arc there is a high arc voltage and hence lower current so a lower burn-off rate, which allows the arc to shorten, enabling current and burn-off rate to return to normal. With a short arc there is a low arc voltage and hence a higher current and a higher burn-off rate which allows the arc to return to normal.

Normally approx. one third parent metal is fused per unit volume of filler wire ensuring a very economic process but only in the flat (or horizontal and vertical position for fillets). A.c. and d.c. power sources may be used with current ranges from 300 to 2000 amps with deposition rates varying between 7 and 45 kg/h using single or multiple wires. The highest deposition rates are with electrode negative but maximum penetration is with the electrode positive. Single wire d.c. is used when fast accurate starts and close arc control are necessary and difficult contours at maximum speed are to be followed with close control over bead shape.

Advantages and Applications

High deposition rates, no visible arc with little fume or spatter, continuously smooth weld surface for long lengths. Sound weld metal with few defects. Deep penetrating pattern. Multi-pass welds in thick sections have refining effect on structure.

Electro-Slag Welding

Electro-slag welding is a process of welding in which the welding heat is generated in the liquid slag bath by the I^2R power dissipation in the slag layer (where I = current and R = resistance), producing a high slag temperature of the order of 1750 to 2000°C. The slag becomes electrically conductive at about 1000°C. There is no arc, the wire melting off as it is fed into the slag pool. Fusion of the parent plate then takes place and the molten metal, which is contained by the copper shoes, solidifies as the carriage, electrodes and shoes all move vertically upwards, leaving the solidified weld behind. (See Fig. 20.2.)

The plate material which is edge prepared to a square edge, either by planing or flame cutting, is set up in the vertical position with a parallel gap of approximately 25 to 50 mm depending upon plate thickness. Water-cooled copper "shoes" at the front and back of this parallel gap thus form a rectangular mould in which melting takes place.

A starting block is welded on at the bottom of the groove and an arc is struck with either one or more continuous

Fig. 20.2 Electro-slag welding
1 Plates to be welded
2 Shoes
3 Molten slag
4 Electrode
5 Molten metal
6 Finished weld
7 Pipes for cooling medium

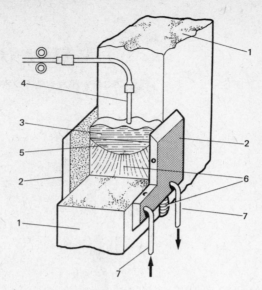

electrodes (depending on material thickness). The arc is struck under powdered flux and is maintained until sufficient liquid slag is produced (30 to 50 mm in depth) when the arc becomes extinguished. At this point the current is raised and the voltage lowered and the process then becomes full electro-slag welding.

The power source may be a.c. or d.c. with current ranges between 400 to 1500 amps. Welding speed is dependent upon metal thickness and varies between one and three metres per hour (50 to 250 mm thick). Due to the very large grain size which forms in the weld during cooling, a normalising heat treatment is often carried out at approximately 920°C for steels to restore weld toughness.

Advantages and Applications

1 High joint completion speed for thick sections.
2 Little distortion of the fabrication.
3 Sound weld metal with few defects.
4 Economical plate edge preparation.
5 Thinner and longer sections in shipyards. Used on thick-walled pressure vessels and steel mill handling equipment.

21 Consumable Guide Welding

The basic principles of the consumable guide process can be seen from Fig. 21.1. A tube, which is coated with slag forming and alloying elements, is used to guide a wire from a feed unit into a bath, formed by the two sides of the joint and two water-cooled copper shoes. This bath contains the molten metal and slag. The tube (i.e. guide) is connected to a power source (positive pole) which in this case is a rectifier. The heat necessary to melt the guide, the filler wire and the joint edges is generated by the passage of the welding current through the ionised slag bath so that the guidetube and wire melt at a rate which determines the welding speed.

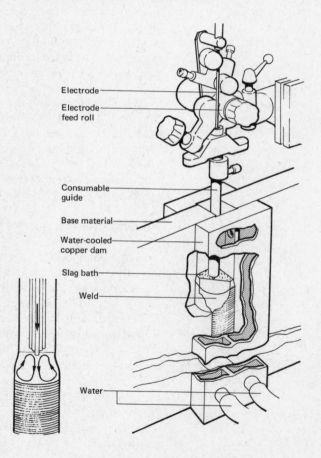

Electrode

Electrode feed roll

Consumable guide

Base material

Water-cooled copper dam

Slag bath

Weld

Water

Fig. 21.1

Due to resistance heating the length of the tube is limited to approximately 1000 mm but this can be extended if a movable current contact is used which enables the current to be fed to the guide at intermediate positions not exceeding 900 mm from the weld pool.

Compared with traditional electroslag welding the obvious advantages are:

a)   The ease of operation in using a lightweight machine.
b)   The suitability for welding thinner plates.
c)   The increase in welding speed.

The increase in welding speed is mainly due to the reduction in joint volume by the use of a consumable guide for feeding the welding wire.

Increased welding speed is obviously advantageous from the point of view of productivity but has the additional advantage that the mechanical properties of the weld metal and the heat affected zones are also improved.

22   Resistance Welding (Spot and Seam)

Resistance welding is welding by electrical heating combined with pressure.

Spot Welding

Spot welding is a form of resistance welding in which a weld is produced at a spot in the workpiece between current-carrying electrodes, the weld being of approximately the same area as the electrode tips, or as the smaller of tips of differing size. Force is applied to the spot, usually through the electrodes, continuously throughout the process. (*No arc is formed.*)

Important Points
1   The copper alloy electrodes have less resistance to the flow of electricity than the material being welded.
2   The greater the resistance to current flow, the more concentrated the heating effect.
Note: Small resistance $R_1 + R_2$ (Fig. 22.1) is between the electrodes and the plate, and high resistance R_3 is the small air gap between the plates, causing a concentrated heat build-up which results in melting and fusion of the two surfaces in the form of a weld nugget.

3 The weld sequence is as follows:
a) the force is applied through the electrodes;
b) current flows for a set period of time;
c) the electrode force is maintained for a period after the current stops flowing, then is released.

4 The heat source is obtained from a TRANSFORMER which supplies the welding electrodes with a HIGH current and a very LOW voltage.

Seam Welding

Seam welding is a form of resistance welding in which force is applied continuously and current intermittently to produce a linear weld, the workpiece being between two electrode wheels or between an electrode wheel and an electrode bar (Fig. 22.2). The wheels apply the force and current and rotate continuously during the making of a linear weld.

Important Points
1 One roll is usually power driven at a set speed.
2 The welding current is passed through the electrode wheels intermittently (on and off continuously) to produce overlapping pressure tight welds.
3 This process is very fast.

Welding Investigation 14
Distortion and Thermal Joining Processes

THEORY Distortion of welded work, i.e. buckling, twisting, bowing or moving out of alignment, is caused by stresses set up in the metal by the heat of the joining process. The more heat the greater the stress and, if the stress is not able to relieve itself either by the metal being allowed to expand or contract freely, then the greater the amount of distortion.

EQUIPMENT T.A.G.S., M.A.G.S. and m.m.a. welding plant, oxy-acetylene torch, resistance spot welding plant, 12 pieces of 150 mm × 50 mm × 1.6 mm level steel plate, 1.6 filler, brazing wire and flux, 2.55 mm dia. electrode.

PROCEDURE
Butt weld all the plates on the long side using each process, excepting the pair brazed and the pair with a 6 mm lap which will be spot welded. Use the minimum amount of heat to produce the best joint. Draw and complete the table.

QUESTIONS
1) Comment on *a*) the process which caused the most distortion and outline the causes and *b*) the process which caused the least distortion and outline the causes.
2) How would using 12 mm thick plate alter the above results?

Fig. 22.1 Spot welding

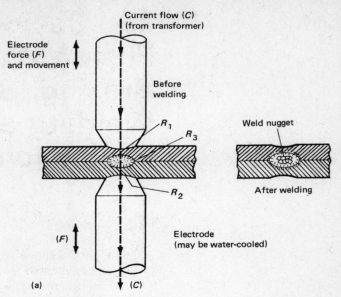

Current flow (*C*)
(from transformer)

Electrode force (*F*) and movement

Before welding

R_1

R_3

R_2

Weld nugget

After welding

Electrode (may be water-cooled)

(a)   (*C*)

Fig. 22.2 Seam welding

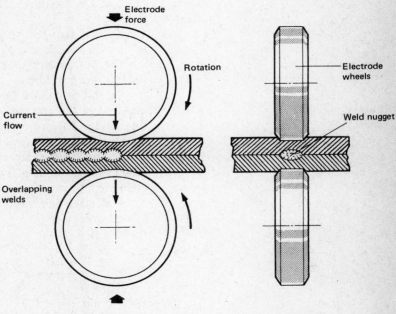

Electrode force

Rotation

Current flow

Electrode wheels

Weld nugget

Overlapping welds

SIDE ELEVATION   END ELEVATION

Final length	Width	Welding time	Angular displacement	Bow	Twist	Process
						M.A.G.S.
						T.A.G.S.
						Oxy-acet. (weld)
						Oxy-acet. (braze)
						Stick weld
						Spot weld (resistance) every 12 mm

23 Basic Principles of Fusion Welding, Brazing, Bronze Welding and Solid Phase Joining

Fusion Welding

A method of welding in which the weld is made between metals in a molten state without the application of pressure.

The important points regarding fusion welding using METALLIC ARC or OXY-ACETYLENE processes are as follows:

1 A very high temperature heat source is required to *melt* the *parent metals* and the *filler* when used.

2 The hot, molten metal must be protected from the atmosphere during welding.

3 There is a change of the physical properties of the welded material in the zone where it has been affected by the heat. Fig. 23.1 is for normalised low carbon steel.

4 Due to the high input of heat, a fair amount of expansion, then contraction, takes place which may result in alteration of dimensions, buckling, twisting, bowing and in severe cases, cracking.

5 A neutral flame is always used for oxy-acetylene welding low carbon steel.

Fig. 23.1

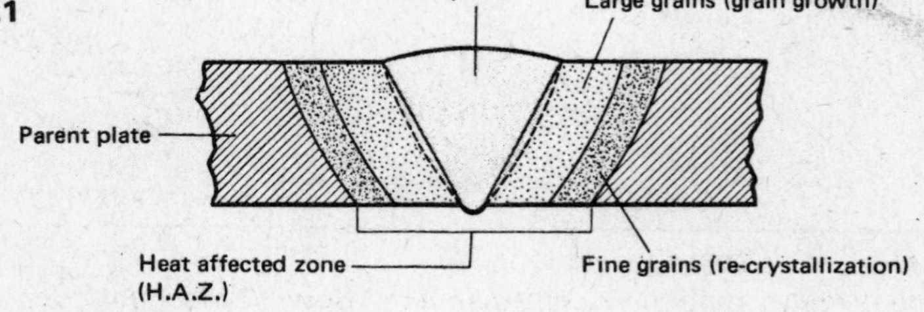

Columnar crystals

Large grains (grain growth)

Parent plate

Heat affected zone (H.A.Z.)

Fine grains (re-crystallization)

Brazing (Hard Soldering)

A process of joining metals in which, during or after heating, molten filler metal is drawn by capillary attraction into the space between closely adjacent surfaces of the parts to be joined. In general, the melting point of the filler metal is above

500°C, but ALWAYS below the melting point of the parent metal.

The important points regarding flame brazing are as follows:

1 Because the parent metals are *not* melted, lower temperature flames can be used, i.e. air coal gas.

2 A flux must be used to clean the surfaces of the joints chemically (remove oxide *not* grease, rust or dirt) to prevent atmospheric oxidation and to reduce impurities or float them to the surface.

3 "Wetting" of the joint must take place to enable the filler to bond and surface alloy with the material being joined.

4 There is an optimum gap where capillary attraction is greatest.

5 Dissimilar metals are readily joined.

6 An oxidising flame is used to prevent zinc loss when brazing, particularly when using spelter filler rod (50% zinc + 50% copper).

Bronze Welding

The joining of metals using a technique similar to fusion welding, using a copper-rich filler metal with a lower melting point than the parent metal, but *neither* using capillary action as in brazing nor intentionally melting the parent metal.

Important points regarding bronze welding are as follows:

1 The filler metal used for bronze welding consists basically of copper and zinc. It may also contain nickel, manganese, silicon or other metals.

2 Some type of flux must be used, commonly a borax type, which serves the same purpose as in brazing.

3 Again, "wetting" plays an important part in this process.

4 Dissimilar metals, copper and case iron, are readily joined.

Solid Phase Joining (or Welding)

This is a welding process in which a weld is made by a sufficient pressure to cause plastic flow of the surfaces, which may or may not be heated.

The important points regarding solid phase joining are as follows:

1 No flux is used, but the metals to be joined must be reasonably chemically clean during bonding.

2 No filler metal is required.

3 Dissimilar metals are readily joined, such as copper and aluminium.

QUESTIONS

1) Put the following processes in a column, the one with the highest heat source first and the one with the lowest last: Bronze welding, Brazing, Oxy-acetylene welding, Manual metal arc welding.

2) Which two gases, from the atmosphere, must be prevented from entering the weld metal?

3) Which of the processes in Question 1 causes a great change in the physical properties of the parent metal?

4) Heating to a high temperature for a period of time causes the grains to:

a) became larger *b*) become smaller *c*) remain the same *d*) elongate.

5) Another name for buckling and bowing due to the heat from welding is:

a) distortion *b*) shrinkage *c*) contraction *d*) overheat.

6) Wide/Narrow gaps are advised for capillary attraction.

7) State two reasons for your answer in Question 6.

8) Brazing uses both capillary and "wetting" actions, but bronze welding uses only one of these. Which?

9) A brazing or bronze welding filler must

a) melt at a temperature above the melting point of the parent metal;

b) melt at a temperature below the melting point of the parent metal;

c) melt at the same temperature as the parent metal.

10) Spelter, which may be used as a filler for both brazing and bronze welding, is composed approximately equally of which substances?

24 The Technique of Hard Soldering, Brazing and Bronze Welding

Hard Soldering and Brazing

A larger flame should be used than for fusion welding with the envelope and not the cone in contact with the work, so that the flame covers a large area. A continual forward and backward movement should be used. The flame should be neutral except when the parent or filler metal contains an appreciable amount of zinc, in which case the flame should be sufficiently oxidising to prevent the zinc fuming off.

The mating surfaces of the metals to be joined should be brought to just above the melting point of the filler, when the flux will boil and melt into the joint. The filler rod tipped with flux should now be touched on to the joint at the hottest part until it begins to melt and flow into the joint. On no account should the rod be melted off and allowed to drop off on to the joint. The joint should now be progressively heated and brazed.

Overheating should be avoided as this weakens the final joint and is recognised by black areas of burnt flux. For joints with thin to thick metals or high and low conductivity metals, the bulk of the heat should be in the thick and high conductivity metal.

Bronze Welding

The surfaces to be joined are painted with a borax type flux and a slightly oxidising flame is used. The edges are then heated up at one point to the flow temperature of the filler rod (850–900°C). The flux-tipped rod is now melted on to the heated surface. The filler wets the joint surfaces and tinning of the surfaces takes place and forms a bond between the filler and parent metal. The wetted area must always be kept in front of the deposited filler and overheating avoided.

Welding Investigation 15
Hard Soldering, Brazing and Bronze Welding

AIM Application of hard (silver) soldering, brazing and bronze welding.

EQUIPMENT AND MATERIALS Natural gas, air blow-pipe, oxy-acetylene welding equipment, zinc-copper (spelter) rod, borax type flux, copper-silicon (bronze) welding rod, silver solder rod and flux. (See Tables 24.1 and 24.2) Metal squares 40 mm square × 1.6 mm, seven of copper, five of each metal, brass, low carbon steel, galvanized steel, two pieces of 90° bevelled grey cast iron, emery cloth, pliers.

PROCEDURE
1) Using the oxy-acetylene leftward technique, bronze weld together the two pieces of cast iron to form a single vee butt joint.
2) Clean all surfaces with emery. Mix a small quantity of brazing flux with water and coat the laps. Lap together for 10 mm as shown in Fig. 24.1 and ensure a small gap approximately 0.08 mm with the sheared edge upwards (a lap of 3 times sheet thickness is usually used). Braze using spelter tipped with flux and the air natural gas torch (note no colour change on some of the metals), in the order shown in Fig. 24.1.
   Alternatively an oxy-acetylene torch may be used and welding rods selected from Table 24.2.
3) Clean surfaces of remaining two copper laps and mix a small quantity of silver solder flux and coat for a distance of 20 mm. Lap for 20 mm and place in close contact (see

Fig. 24.1). Hard solder using air gas torch with silver solder tipped with flux.

A comparison may be made by using the oxy-acetylene torch and the bronze rod to join dissimilar strips.

4) Cut each joint with a saw as shown in Fig. 24.1a and file and polish smooth to show the length of capillary flow of the filler (use low powered lens) and show in red on the sections. Using pliers peel back one of the half strips on each joint to test the bond.

Place the cast iron in the vice and attempt to break along the length of the bronze weld (care).

A saw cut down the centre of the weld and at the ends will assist the fracture.

QUESTIONS

1) Which filler was the easiest flowing?
2) Which joint was the easiest to make?
3) Name the filler which has the lowest melting point?
4) What is the name of the white deposit and fumes off the galvanized plate?
5) After brazing, which metal became softest and easy to bend?
6) Bend a piece of spelter rod and a piece of bronze rod and say which is the most ductile.
7) List those materials used for which it was difficult to judge the colour change and hence the temperature.
8) Which metal fractured: the cast iron, the bronze or the bond? Give reasons.

25 Principles of Soft Soldering

Solder is a metallic substance used for joining metals (see Table 25.1, for various compositions).

The basic principles are as follows:

1 The melting point of the solder is *always* lower than the melting point of the metals being joined.

2 The surfaces of the metals to be joined must be clean of dirt and grease and paint.

3 A flux must be used to remove the film of oxide and to promote wetting of the surfaces by the tin in the solder.

4 The liquid solder must displace the flux (Fig. 25.1a) so that the tin reacts with the metals being joined and so forms a compound (alloy) which is extremely strong (Fig. 25.1b).

Fig. 24.1

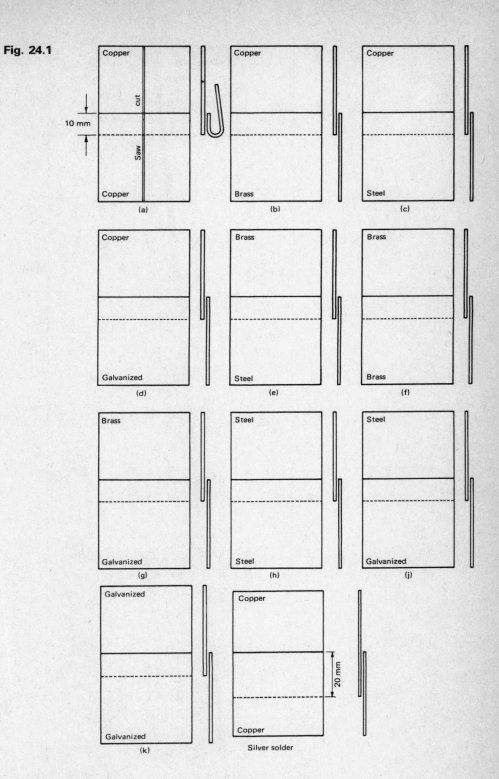

Table 24.1 Filler Rods and Fluxes for Silver Soldering, and Brazing

BS. 1845 Type	Approx. composition %			Other Elements	Type of Flux	Suitable Parent Metals	Joint Clearance (mm)	Melting Range °C	Remarks and Use	
	Silver	Copper	Zinc							
Ag. 1	50	15	16	Cadmium 19	Chlorides + Fluorides	Brass Nickel	0.003	620–640	Low melting point	Silver solders Very fluid Small gaps
AG. 4	61	28	10		Borates	Bronze Steel	0.009	690–735	High conductivity	
AG. 5	43	37	19			Copper	0.009	700–775	For general engineering	
CP. 1	14	Remainder		Phosphorus 5	Borates + Fluorides	Copper	0.009 to 0.015	645–700	Not used on nickel or ferrous alloys	
CP. 3		Remainder		Phosphorus 7	Fluorborates			705–800		
CZ. 1		50	Remainder		Borax + Boric acid	Steel Copper	0.006 to 0.015	860–870	Spelter-brazing	Brasses
CZ. 2		54	Remainder			Brass		870–880	For general engineering purposes	
CZ. 3		60	Remainder			Galvanized		885–890		

Table 24.2 Bronze Welding and Brazing

BS. 1453	Copper	Nickel	Silicon	Manganese	Iron	Suitable Parent Metals	Melting Range	Remarks
C.2	60		0.3			Copper and low carbon steel	All above 850°C	All elements are taken to the mean average Remainder zinc
C.4	60		0.2	0.2	0.3	Galvanized		
C.5	50	10	0.3	0.5	0.5	Cast and malleable iron		
C.6	43	12	0.3	0.2	0.3	Combination of any two of above		

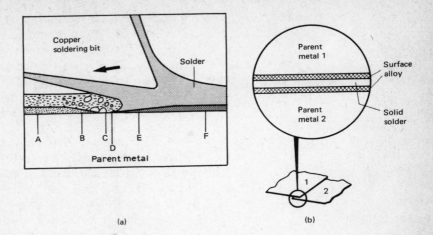

Fig. 25.1

A   Flux solution lying above oxidised metal surface
B   Boiling flux solution removing the film of oxide
C   Bare metal in contact with fused flux
D   Liquid solder displacing fuxed flux
E   Tin reacting with the basis metal to form compound
F   Solder solidifying

5   The parts to be joined should fit closely so that the space between them is narrow enough to fill with molten solder, drawn in by capillary force as the solder wets the adjacent sides. (A clearance of 0.08 to 0.13 mm is satisfactory for untinned surfaces.)
6   Some form of heat must be employed to heat the metals to be joined and to melt the solder. The two most common methods are copper bit soldering irons and various gas flames.

Fluxes

As mentioned above the purpose of a flux is to remove thin films of oxide and float them away, and to prevent further oxidation of the chemically clean surface.

A flux should have the following characteristics:
1   It should be a liquid cover over the part to be soldered and exclude any oxygen from the atmosphere.
2   It should dissolve any oxide on the metal surface or on the solder and carry away these unwanted materials.
3   It should be readily displaced from the metal required to be joined by the fluid solder.

Basically there are two types of flux used when soft soldering.

1 Corrosive (Active)

The corrosive fluxes are effective and quick acting on a wide variety of metals, i.e. steel, copper, brass, zinc, tin, lead, cadmium. They must only be employed when the flux residue is able to be removed due to its highly corrosive nature.

The corrosive fluxes are named HALIDE FLUXES and contain one of the following acids: fluorine, chlorine, bromine or iodine.

A most common flux is zinc chloride (see below). This is manufactured by adding small granules of zinc to dilute hydrochloric acid. The reaction which follows produces heat and gives off hydrogen gas, leaving behind zinc chloride flux. Traces of this flux must always be removed after soldering by washing in hot water to prevent corrosion.

The highly corrosive zinc chloride is banned on electrical work or work which cannot be effectively washed.

2 Non-Corrosive (Passive)

These fluxes are of the Resin and Tallow type. Resin is the gum found in pine trees and is only active at the soldering temperature, leaving a residue which is non-corrosive and even protective. Resin flux is used for electrical work and delicate instruments and pressure gauges, where washing is difficult. Resin flux is suitable for tin plate and hot-dipped tinned metals, but is too slow acting for steel and untinned metals. Flux cored solder has a core of resin with a slight addition of acid. This is called an activated flux.

Welding Investigation 16
Preparation of Zinc Chloride Flux

AIM To prepare a zinc chloride flux known as killed spirits (corrosive).

EQUIPMENT 500 cc beaker, 200 cc measuring cylinder, zinc granulated or snippings, hydrochloric acid, 38% strength, sal ammoniac, distilled water, glass rod, four test tubes and corks, funnel. (See Fig. 25.2.)

SAFETY *Always* ADD ACID TO WATER, *never* the reverse.

This will prevent the acid reacting violently and splashing about.

Do not allow acid to come in contact with eyes, skin or clothes as it can cause quite severe burning and damage.

PROCEDURE

Measure 100 cc of distilled water and pour into beaker. Now measure 10 cc of 38% hydrochloric acid and add carefully to water, stirring constantly. Add the zinc chips, a few at a time, to the liquid in the beaker and note the reaction of acid on zinc. Collect the gas given off in the test tubes and cork.

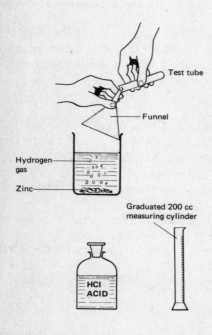

Fig. 25.2 Apparatus for preparing zinc chloride

At the end of the reaction, there should be a surplus of zinc at the bottom along with some impurities from the zinc such as lead. Add a little sal ammoniac to ensure that all the acid has been "killed".

Test the samples of gas given off by uncorking and applying a lighted splint to the open end and note the result.

Note: This flux must not be used for electrical wiring because of its corrosive nature.

Now draw and complete a table as shown.

Flux produced	Acid used	Metal used	Gas given off	Does gas explode?	Another name for flux

QUESTIONS

1) Never/Always add water to acid.
2) Name the substance formed when hydrogen gas reacts with air.
3) The gas given off when zinc is added to hydrochloric acid is
a) Oxygen, b) Hydrogen, c) Nitrogen, d) Chlorine.
4) Hydrochloric acid would be stored in a glass vessel carrying the following formula:
a) CIH, b) HCl, c) H_2SO_4, d) HFl
5) An active flux should *not* be used on electrical assemblies for the following reasons:
a) It may burn the fingers.
b) It may cause fire.
c) It corrodes the joint.
d) It gives off an explosive gas.
6) The proportion of 38% hydrochloric acid to water is
a) 100 to 1, b) 10 to 1, c) 500 to 1, d) 200 to 1
7) If acid splashed on the hands or face, you should
a) immediately wash it off
b) run to the First Aid Room
c) wipe it off with a cloth
d) report to the lecturer

Table 25.1 Solders for General Use (British Standard Specification BS 219)

B.S. Solder	Tin Content Max. %	Min. %	Antimony Max. %	Min. %	Lead %	Typical Applications
A	65	64	0.6	—	remainder	Components liable to damage by
K	60	59	0.5	—	remainder	heat or requiring free-running
						solder, e.g. electrical, radio and
						instrument assemblies, machine
						soldering of can ends.
B*	50	49	3.0	2.5	remainder	Coppersmith's and tin-smith's bit
F	50	49	0.5	—	remainder	soldering; general machine soldering,
M*	45	44	2.7	2.2	remainder	e.g. can-end seams.
R	45	44	0.4	—	remainder	
C*	40	39	2.4	2.0	remainder	Blow-pipe soldering, soldering of
G	40	39	0.4	—	remainder	side seams on high speed body-forming machines.
H	35	34	0.3	—	remainder	Plumber's solder, wiping of cable
L*	32	31	1.9	1.6	remainder	and lead pipe joints.
D*	30	29	1.8	1.5	remainder	Dipping baths.
J	30	29	0.3	—	remainder	
V	20	19	0.2	—	remainder	Electric lamp manufacture,
N*	18.5	18	1.1	0.9	remainder	dipping solder.

* Antimonial, not recommended for zinc and galvanized work.

26 Capillarity

Capillary action or force plays a very important part in the soldering and brazing processes and is the result of the molecular attraction between the liquid and the chemically clean solid. The maximum capillary "climb" of the liquid is obtained only when the liquid effectively wets the solid and when the gap between the components is of the optimum distance, usually only a few hundredths of a millimetre.

Note: The surface tension of the liquid at the edge pulls the liquid up against the pull of gravity.

An example of molecular attraction of a liquid for a solid is shown when an open-ended fine bore glass tube is placed in water. The level of the water in the tube rises above the normal water level in response to the attraction of molecules on the surface of the chemically clean glass. If, however, oil or grease is present around the bore, no climb takes place, as there is no molecular attraction between oil and water.

When clean low carbon steel is being soldered or brazed, the tin or the copper will wet the steel and creep upward just as the water did on the clean glass. If the steel surface is covered with grease, dirt, burnt flux or thick oxide, then there is no molecular attraction between the tin or copper for the steel and it behaves like the water and grease.

Wetting occurs when the molecules of a liquid adhesive are attracted by the molecular force of the chemically clean solid and spread to form an immensely strong bond.

Welding Demonstration 6
Gap Size and Surface Condition in Capillarity

AIM   To show the effect of gap size and surface condition on liquid climb.

EQUIPMENT   Glass trough, thermo-chromic liquid or meths, four bulldog clips, metal shims (two 0.025 mm, two 0.25 mm, one 0.5 mm), degreasing agent, eight pieces of clear welding glass.

Fig. 26.1

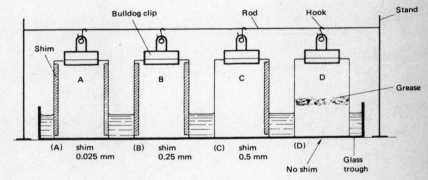

PROCEDURE

Degrease the glass plates and clip together in pairs as in Fig. 26.1. Insert the shims between glass to obtain a parallel gap for A and B and a glass to glass contact along one edge for C, and for D a band of grease and close contact without a shim.

Hook the clips on the bar as shown and leave upright for two minutes, then draw in the various levels of the coloured liquid in red, and complete the Table.

	Did wetting take place?	Max. height climb (mm)	Min. height climb (mm)	Additional remarks
A				
B				
C				
D				

Welding Demonstration 7
Effect of Bore Size on Liquid Climb

AIM To show the effect of bore size on liquid climb.

EQUIPMENT Glass dish, capillary tubes of varying bores, water plus methyl orange, tube holder plus rubber corks. See Fig. 26.2.

PROCEDURE

Place cleaned tubes into liquid with tube with smallest diameter to largest diameter in ascending order, and then, using a red colour, mark in the relative heights of the liquid "climb" and complete the Table.

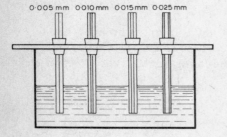

0.005 mm   0.010 mm   0.015 mm   0.025 mm

Fig. 26.2

Bore	Diameter (mm)	Height of Climb (mm)
A		
B		
C		
D		

Welding Demonstration 8
Effect of Surface Condition in Soldering

AIM To show the effect of the surface condition of the material being soldered on its wettability.

EQUIPMENT Four pieces of 50 mm square 20 s.w.g. copper sheet, tripod stand, bunsen burner, zinc chloride flux, emery cloth, flint lighter, solder grade F, B.S. 219 (page 114).

PROCEDURE

Set up the equipment as in Fig. 26.3 and prepare the copper sheet as follows:

a) Leave in the tarnished state.

b) Leave in tarnished state but add some drops of zinc chloride flux in the centre.

c) Emery cloth thoroughly all over.

d) Emery cloth thoroughly and add a few drops of zinc chloride flux.

Place a pea-sized piece of solder in the centre of each square and heat from below until solder melts and note the results. Saw through centre of each square and try to break the bond between solder and copper.

Complete the Table after sketching the results.

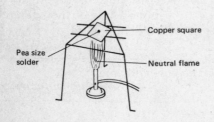

Copper square

Pea size solder

Neutral flame

Fig. 26.3

Surface condition	Size Spread (mm) Length	Breadth	Height Cross-section	Bonding ability*
A				
B				
C				
D				

* Excellent, good, poor.

QUESTIONS

1) A liquid will reach its maximum climb if
a) there is no gap;
b) there is a very small gap;
c) there is a medium gap;
d) there is a large gap.

2) Molecular attraction of a liquid for a solid is assisted by
a) a light film of grease;
b) a tarnished surface;
c) a slight oxide film;
d) a bright clean surface.

3) Tick those of the following which assist tin to wet a metal surface:
a) zinc chloride flux;
b) tin oxide;
c) heat;
d) water.

4) If wetting ability is measured by the spread of solder, list in order, best first, the surface conditions which allowed wetting.

27 Material Removal by the Oxy-Fuel Process

Manual Cutting

Steel, when heated to a bright red (approximately 875°C), will oxidise rapidly (burn) when a jet of high pressure oxygen is directed on to it. This principle is utilised for cutting steel plates and sections.

A special blowpipe and nozzle is used with higher oxygen pressures than those used for welding. A burning mixture of the fuel gas and oxygen is used to pre-heat the steel to ignition temperature, then a lever is used to open a valve which releases the high pressure cutting oxygen which blows away the fluid iron oxide formed, leaving a narrow cut called the *kerf*. The nozzle is moved progressively at a constant height (called stand-off distance) and speed in the direction of cutting. The nozzle should always be at 90° to the surface unless bevel cutting is being carried out. The tips of the blue cones should be kept at about 3 to 5 mm above the plate surface.

Flame Adjustment

Pre-heating prior to cutting is always carried out with a flame adjusted to neutral (Fig. 27.1).

Fuel Gases

Table 27.1 shows the main fuel gases used in industry.

Table 27.1 Fuel Gases

Fuel Gas Chemical formula	Method of Production and Storing	Colour of Cylinder	Approx. Flame Temp. with Oxygen	Applications and Advantages
Acetylene C_2H_2	Dissolved in acetone in cylinders of porous mass. Calcium carbide+water	Maroon	3100°C	Welding. Brazing, bronze-welding, cutting thick sections. Pre-heating quick starting. Highest cal. value.
Propane (L.P.G.) C_3H_8	(Liquid petroleum gas) In cylinders or bulk. "reforming" of paraffins.	Red	2820°C	Liquid bulk. Clean cuts. Pre-heating. Brazing, bronze welding thick sections. Little soot. Popular.
Natural Gas (L.N.G.) $CH_4+C_2+H_6$	North Sea gas (refrigerated liquid gas). High pressure pipeline.	Red	2800°C	Brazing. Bronze welding. Cutting. Pre-heating. Soldering.
Hydrogen H_2	High pressure gas in cylinders. "Reforming" of oil.	Red	2830°C	Under-water cutting. Brazing and bronze welding. Soldering (furnace).
Methane CH_4	High pressure gas in cylinders. North Sea and decomposition of sewage.	Red	2770°C	Brazing. Pre-heating. Cutting.
Intermediate gas mixtures. *Trade names* M.A.P.P. APACHI	Mixtures of several hydrocarbon gases. Liquified in bulk or cylinders.	Red	2920°C	Cutting. Pre-heating. Brazing. Safer than acetylene.

Fig. 27.1

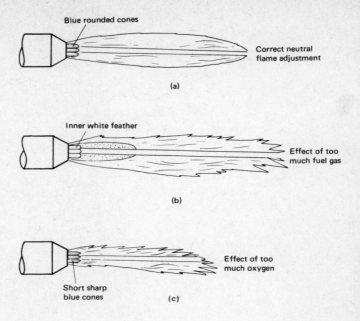

(a)

(b)

(c)

Types of Nozzle

There are several types of nozzle, but all are designed on the same principle, i.e. a central orifice which carries the high-purity high-pressure cutting oxygen and surrounded by one or more gas ports carrying a mixture of fuel gas and oxygen.

The central orifice diameter is always stamped clearly on some part of the nozzle as this is most important when selecting a suitable nozzle for the plate thickness. The thicker the plate, the larger the diameter of the central orifice and also the wider the kerf.

Nozzles fall into two categories: those which mix the pre-heat gases in the nozzle and those which have the gases mixed previously in the blowpipe.

Figs. 27.2, 27.3 and 27.4 show a selection of nozzle types.

Hand-operated cutting blowpipes may be welding blowpipe shanks fitted with a special cutting head or they may be designed solely for cutting. Fig. 27.5 shows an example.

Profile Cutting Templates

From a production point of view, that is cost, quantity and quality, hand cutting is no match for the small semi-automatic profiling machine, especially one with attachments for automatic or hand-guided followers, which trace out the shape of previously prepared patterns. The previously lit cutter head cuts out the pattern shape simultaneously at some distance from the template.

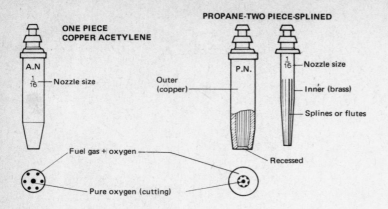

ONE PIECE
COPPER ACETYLENE

PROPANE-TWO PIECE-SPLINED

A.N $\frac{1}{16}$ — Nozzle size

P.N.

$\frac{1}{16}$ — Nozzle size

Outer (copper)

Inner (brass)

Splines or flutes

Recessed

Fuel gas + oxygen

Pure oxygen (cutting)

Fig. 27.2 Nozzle mix types

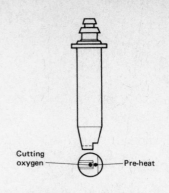

Cutting oxygen — Pre-heat

Fig. 27.4 Step nozzle for sheetmetal (acetylene)

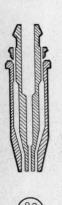  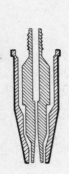 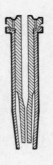 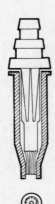 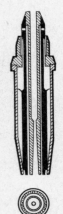

Acetylene
3 seats
1 piece
Parallel bore
Preheat holes
No skirt

Intermediate gas
3 seats
2 piece
Parallel bore
Preheat flutes
Short skirt

Acetylene
2 seats
2 piece
Venturi bore
Preheat annulus
No skirt

Natural gas
2 seats
2 piece
Venturi bore
Preheat flutes
Long skirt

Propane
3 seats
2 piece
Parallel bore
Preheat slots
Long skirt
Cold formed

Propane
3 seats
2 piece
Parallel bore
Preheat flutes
Long skirt
Oxygen curtain

Fig. 27.3 B.O.C. designed nozzles

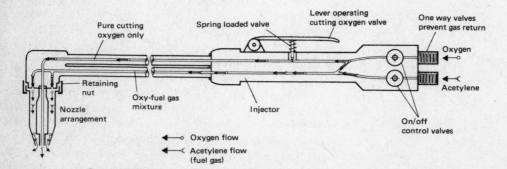

Pure cutting oxygen only

Spring loaded valve

Lever operating cutting oxygen valve

One way valves prevent gas return

Oxygen

Retaining nut

Oxy-fuel gas mixture

Injector

Acetylene

Nozzle arrangement

On/off control valves

Oxygen flow

Acetylene flow (fuel gas)

Fig. 27.5 Principle of the cutting blowpipe

There are several types of profile follower but a selection of the most common only is given.

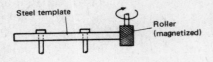

Fig. 27.6

1 Magnetic (Fig. 27.6)
A steel template is clamped rigid and an electronically magnetized roller is driven round the profile of the template. Allowance must be made on template for roller and kerf.

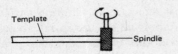

Fig. 27.7

2 Knurled Spindle (Fig. 27.7)
A wood or hardboard template is fastened to the table and the spindle rotating at a set speed is guided by hand around the profile.

Fig. 27.8

3 Serrated Wheel (Fig. 27.8)
A drawing on straw paper or timber is followed by a serrated wheel (or a fixed pointer) revolving at a set speed and guided by hand.

Fig. 27.9

4 Cross-Light (Magic Eye) (Fig. 27.9)
A black line drawing is followed by two beams of light focussed on one spot (the black line) and relay information back to the electronic guiding device.

On outside dimensions MINUS half the roller diameter and ADD half the kerf all round.

On inside dimensions, ADD half the roller diameter and MINUS half the kerf all round.

Example
Required finished size of plate:
Outside radius = 100 mm   Kerf width 2 mm
Inside radius = 50 mm   Roller diameter 6 mm
1 Outside radius minus half roller diameter, add half kerf:

$$(100 \text{ mm} - 3 \text{ mm}) + 1 \text{ mm}$$
$$= 97 + 1 = 98 \text{ mm (outside radius of pattern)}$$

2 Inside radius add half roller diameter, minus half kerf:

$$(50 \text{ mm} + 3 \text{ mm}) - 1 \text{ mm}$$
$$= 53 - 1 = 52 \text{ mm (inside radius of pattern)}$$

Machine Flame Cutting

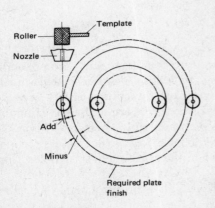

Fig. 27.10 Example showing kerf and roller allowance on a template for magnetic and spindle heads

There are certain factors which influence the accuracy of the cut profile which are common to both manual and machine cut components. The main ones are

1 Rust, millscale and dirt on the plate surface.

2 Incorrect flame adjustment due to incorrect pressures or fouled gas ports.

3 Incorrect procedure (see page 124), effects of variation in procedure.

4 Finally, even when the above factors have been corrected, one other factor is often overlooked, and that is *inaccuracies due to distortion.*

The heat from cutting causes expansion stresses which cause the separated plate and scrap to move relative to each other, long narrow profiles being most affected. In addition surface buckling may develop, necessitating automatic nozzle height control. The greatest distortion occurs in the material having the smallest surface area. For example if a 150 mm disc is cut from a one metre square plate, then the hole would be accurate since the larger area of plate remains stable, and the smaller disc of scrap would be distorted due to the sideways movement of the disc. If, however, a large circular disc of 990 mm diameter is cut from a one metre plate, then the disc would be accurate and the scrap cut off would distort due to sideways movement. Fig. 27.11 shows displacement and kerf closure causing disc error.

Fig. 27.11 Slit closure and displacement of disc gives error

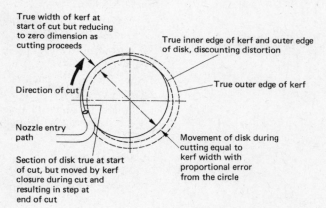

True width of kerf at start of cut but reducing to zero dimension as cutting proceeds

True inner edge of kerf and outer edge of disk, discounting distortion

Direction of cut

True outer edge of kerf

Nozzle entry path

Movement of disk during cutting equal to kerf width with proportional error from the circle

Section of disk true at start of cut, but moved by kerf closure during cut and resulting in step at end of cut

To utilise plate and produce minimum scrap, components are often "nested", that is fitted close together so that wherever possible only sufficient space for cutting is left between components. Often bridge or tie pieces are utilised to minimise distortion and any outwards movement. These may be in the form of sections of uncut kerf or pieces welded across sections which are later cut out when the component has cooled (Fig. 27.12).

Zig-zag entry and small wedges also control plate movement (Fig. 27.13).

Distortion due to the cutting heat may be minimised by sequential cutting which enables the scrap to move and allows the required profiles to remain stable.

Fig. 27.12

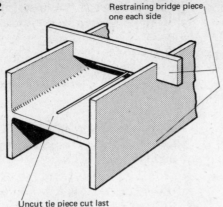

Restraining bridge piece
one each side

Uncut tie piece cut last

Fig. 27.13

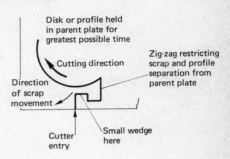

Disk or profile held
in parent plate for
greatest possible time

Cutting direction

Zig-zag restricting
scrap and profile
separation from
parent plate

Direction
of scrap
movement

Cutter
entry

Small wedge
here

Cutting Demonstration 1
Oxy-Fuel Gas Cutting

AIM Demonstration of oxy-fuel gas cutting and safe use of equipment.

EQUIPMENT Oxy-fuel gas equipment, cutting blowpipe and nozzle to suit fuel gas and metal thickness. Goggles, heat-resisting gloves, apron, and feet protectors, flint lighter and tongs.

SAFETY Ensure that no combustible material is near and likely to catch fire, and that the cut portion will not fall on to hoses or any person.

PROCEDURE
Demonstrate straight line manual cutting using angle section as a guide and freehand cutting using a spinner arm and freehand, then compare with semi-automatic profiler. Use Questions 1–31 on page 127 as part of the Demonstration.

Effects of Variation in Procedure on Cut Surface

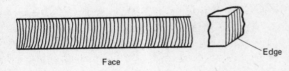

Face

Fig. 27.14

Edge

1 Acceptable Cut (Fig. 27.14)
Cause Correct nozzle height and flame adjustment with correct speed of travel. Correct nozzle size and pressures for plate thickness.

Identification Sharp square top and bottom edges. Vertical drag lines curving slightly at the bottom with no adhering dross. Oxide easily brushed off to give a ripple free face. Any centre dots cut in half.

123

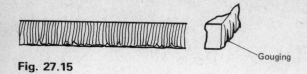

Fig. 27.15

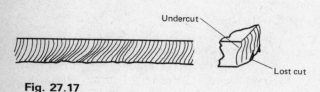

Fig. 27.16

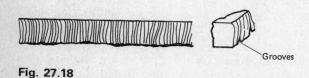

Fig. 27.17

Fig. 27.18

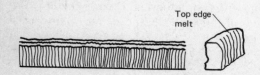

Fig. 27.19

Fig. 27.20

2 Not Enough Pre-heat (Fig. 27.15)
Cause Low pre-heat adjustment results in slow speed of travel, causing oxygen in cutting stream to gouge face.

Identification Gouging and grooves into bottom of cut face.

3 Excessive Pre-heat (Fig. 27.16)
Cause Excess pre-heat adjustment causes melting due to excess heat.

Identification Melted and rounded top edge with rough appearance. Dross on bottom edge difficult to remove and sometimes floods kerf gap.

4 Cutting Speed too Fast (Fig. 27.17)
Cause Incorrect machine speed setting. Too large a nozzle size for plate thickness.

Identification Drag lines slope back excessively. The face may be slightly under-cut and the bottom may not be completely cut in places, in fact the cut may be lost altogether.

5 Cutting Speed too Slow (Fig. 27.18)
Cause Incorrect machine speed setting. Too small a nozzle size for plate thickness.

Identification Melted and rounded top edge. Lower part of cut face has grooves gouged out with a rough bottom edge and adhering dross on bottom edge difficult to remove.

6 Fluctuating Speed of Travel (Fig. 27.19)
Cause Machine tracing wheelskid. Hand cutting without guides.

Identification Uneven drag lines and cut edge occasional adhering dross on bottom edge.

7 Nozzle too High (Fig. 27.20)
Cause Warped plate or incorrect initial stand off distance.

Identification Melted and rounded top edge, maybe some undercutting of top edge.

Cutting Investigation 1
Procedure and Distortion

AIM To note the distortion effects and variation in cutting procedure on cut faces.

EQUIPMENT AND PROCEDURE As for Demonstration 1. Compare your cuts with Figs. 27.14–27.20 and, where they are less than acceptable, state the remedies.

Place the cut faces of the plates against their matching faces, check for distortion, and state the remedy.

Table 27.2 Data Required for Manual Oxy-Fuel Gas Cutting Saffire High Pressure Blowpipes

SAFFIRE A-NM ACETYLENE NOZZLES

Plate Thickness mm	Nozzle Size	Operating Pressures		Gas Consumption			Approx. Cutting Speeds mm/min
		Oxygen bar	Acetylene bar	Cutting Oxygen l/h	Heating Oxygen l/h	Acetylene l/h	
6	$\frac{1}{32}$	1.8	0.14	800	480	400	430
13	$\frac{3}{64}$	2.1	0.21	1900	570	510	360
25	$\frac{1}{16}$	2.8	0.14	4000	540	470	280
50	$\frac{1}{16}$	3.2	0.14	4500	620	560	200
75	$\frac{1}{16}$	3.5	0.14	4800	680	620	200
100	$\frac{5}{64}$	3.2	0.14	6800	850	790	150
150	$\frac{3}{32}$	3.2	0.21	9400	960	850	130
200	$\frac{7}{64}$	4.2	0.28	14 800	1380	1250	100
250	$\frac{1}{8}$	5.3	0.28	21 500	1560	1420	50
300	$\frac{1}{8}$	6.3	0.28	25 000	1560	1420	50

SAFFIRE P-NM PROPANE NOZZLES

Plate Thickness mm	Nozzle Size	Operating Pressures		Gas Consumption			Approx. Cutting Speeds mm/min
		Oxygen bar	Propane bar	Cutting Oxygen l/h	Heating Oxygen l/h	Propane l/h	
6	$\frac{1}{32}$	2.1	0.2	1000	1300	300	430
13	$\frac{3}{64}$	2.1	0.2	1800	1600	300	360
25	$\frac{1}{16}$	2.8	0.2	3900	1700	400	280
50	$\frac{1}{16}$	3.2	0.3	4500	1800	400	205
75	$\frac{1}{16}$	3.5	0.3	4800	2000	500	205
100	$\frac{5}{64}$	3.5	0.3	7300	2600	600	152
150	$\frac{3}{32}$	4.2	0.4	12 300	3300	800	125
200	$\frac{7}{64}$	4.9	0.4	15 800	4200	1000	100
250	$\frac{1}{8}$	5.6	0.6	22 300	4600	1100	50
300	$\frac{1}{8}$	6.7	0.6	26 300	5900	1400	50

A nesting of plates for oxy-fuel gas cutting. Note the plate economy, sequence of cutting, and tee-section support. (*Courtesy B.O.C. Ltd.*)

A multi-nozzle automatic machine cutting of thick plate. Note the pre-heat nozzle for the thick plate, the protective clothing of the operator, the method of supporting the work, and the neatly stacked finished work. (*Courtesy B.O.C. Ltd.*)

28 Oxy-Fuel Gas Gouging (Ferrous Metals)

Theory and Principles

Flame gouging depends upon the fundamental principle of all oxygen cutting processes: that iron and steel, if previously heated to a high enough temperature (known as the Ignition Temperature), will burn or oxidise vigorously when in contact with oxygen.

The success of the flame gouging process depends upon the design of a special nozzle (Fig. 28.1) to deliver a relatively large volume of oxygen at low jet velocity. This, coupled with proper preheat flame distribution and correct manipulation of the gouging blowpipe, will cut a smooth, accurately defined groove in the surface of steel plate. By using different sized nozzles and varying the nozzle angle, the groove can be made the width and depth required.

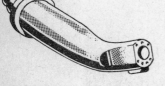

Fig. 28.1 Oxy-acetylene gouging nozzle

Equipment

As for oxy-fuel gas cutting plus special gouging nozzles: A.G. for acetylene and P.G. for propane.

Using a standard blowpipe fitted with the special nozzle shown in Fig. 28.1, the pre-heat is obtained by just touching the inner cones of the pre-heat flames to the work with the torch at an angle of 20°. When the oxygen lever is depressed and gouging begins, the torch angle is lowered to approx. 5° and gouging then progresses. There are two basic gouging techniques: progressive and spot. In Progressive Gouging (Fig. 28.2), a continuous groove is gouged along the plate (the underside of welds). In Spot Gouging (Fig. 28.3) small areas are removed, sometimes to quite a depth.

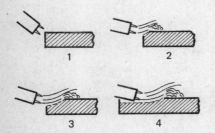

Fig. 28.2 Progressive gouging

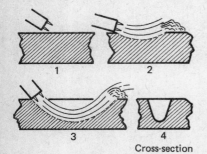

Fig. 28.3 Spot gouging

Advantages and Applications

1 At least three times faster than chipping.
2 – May be used with existing oxy-fuel gas cutting equipment.
3 Useful for removing weld defects, lugs, cleats, tack welds.
4 Dismantling structures and removing risers from castings. Gouging cracks prior to welding.
5 Preparing abutting edges for welding.

QUESTIONS ON CUTTING

1) The ignition temperature of steel is:
a) 575°C b) 875°C
c) 925°C d) 1000°C

2) When a high pressure jet of oxygen is directed on to red hot metal, the following takes place:
a) Melting of the metal.
b) A chemical reaction.
c) An endothermic reaction.
d) Reduction of oxides.

3) A nozzle orifice indicates:
a) The diameter of the bore through the nozzle.
b) The diameter of the bore at the exit point.
c) The diameter of the bore at the inlet point.
d) The general hole size.

4) The relative gas pressures used when fuel-gas cutting are
a) High fuel-gas, low oxygen.
b) Low fuel-gas, low oxygen.
c) Low fuel-gas, high oxygen.
d) High fuel-gas, high oxygen.

5) The roller allowance on outside dimensions for templates is
a) Subtract diameter of roller *overall*.
b) Subtract radius of roller *overall*.
c) Add radius of roller *overall*.
d) Add diameter of roller *overall*.

6) Name the groove produced when fuel-gas cutting plate.

7) Which of the following gases can be supplied as a liquid?
a) Oxygen. b) Propane
c) Acetylene d) Butane

8) Place the following flame temperatures with their respective gases when mixed with oxygen: 3100°C, 2820°C, 2770°C, 2920°C
a) Propane b) Acetylene c) Apachi d) Methane

9) What is the fuel gas used for underwater cutting below 10 m?

10) Name three popular types of nozzle.

11) What is the composition of fuel gases used for welding?

12) State the prime reason acetylene gas has such a high flame temperature with oxygen.

13) All combustible gases have left/right hand threads.

14) Oxide films, scale and laminations have the following effect:
a) Promote oxide fluidity leading to better cutting.
b) Have no significant effect.
c) Promote easier starting of cut.
d) Prevent the production of a first-class cut.

15) Which metals cannot be cut successfully using oxy-fuel gas cutting only?

a) Plain carbon steel b) Copper
c) Stainless steel d) Aluminium

16) What do the "drag" lines indicate on the kerf wall?

a) The oxygen pressure.
b) The speed of the cutter.
c) The nozzle size.
d) The height of the nozzle.

17) Nozzles are manufactured from copper or copper alloy because:

a) It is easy to machine.
b) It is easy to work.
c) It dissipates heat quickly.
d) It does not rust.

18) Which are the fuel gases:

a) Argon   b) CO_2 c) Oxygen
d) Methane   e) Propane f) Nitrogen

19) Oil or grease should never be used on connections carrying high pressure oxygen because of:

a) Explosive risk.
b) Self-lubricating properties.
c) May cause spanner to slip.
d) Dirt collecting with grease.

21) Which flame setting is used for oxy-flame cutting of steel?

a) Oxidising b) Carburising c) Neutral

22) Which cylinders are painted RED?

a) Acetylene b) Coal gas
c) Propane   d) Hydrogen

23) Which connecting nuts have a notch cut on the corner?

a) Oxygen   b) Propane
c) Acetylene d) Coal gas

24) Name four types of profile template followers.

25) A quick accurate method of determining a nozzle size is to:

a) Look at the width of cut.
b) Look on the nozzle.
c) Measure the orifice.
d) Glance at the pressure gauges.

26) Which of the following gases is the cheapest to buy per 1000 cubic metres?

a) Acetylene b) Propane in bulk
c) Coal gas   d) Ferrolene

27) Oxy-fuel gas cutting produces a hard shell at the cut face of low carbon steel mainly because:

a) The oxides are hard.
b) Low carbon steel hardens when quenched.
c) Carbon migrates to the cut edge.
d) Of the great heat.

28) The following metal is never used for conveying acetylene gas due to the potentially explosive compound formed:

a) Copper *b*) Steel *c*) Aluminium *d*) 60/40 Brass

29) The correct procedure to discover the whereabouts of a gas leak would be:

a) Use a lighted match.

b) Use a bucket of water.

c) Use your sense of smell.

d) Use a brush and soapy water.

30) The correct place for a cylinder key during welding or cutting is:

a) In the oxygen cylinder.

b) In the acetylene cylinder.

c) In your pocket.

d) On the workbench.

31) A frozen oxygen regulator should be thawed out by:

a) A hot flame *b*) Hot water

c) Hot metal *d*) A hot fire

32) An acetylene-air mixture is potentially explosive between the following proportions:

a) 0 to 80% *b*) 2 to 82%

c) 10 to 100% *d*) 5 to 95%

29 Oxygen Arc Cutting

Safety

As for other arc processes plus oxygen cutting. Dangers from ejected dross.

Theory and Principles

In the oxy-arc cutting process the heat is provided by an electric arc struck between an electrode and the material to be cut. Due to the intense heat of the electric arc, preheating times are much shorter and cutting speeds higher than in oxy-acetylene cutting. The process has proved particularly advantageous for cutting cast iron and stainless steel. Further applications are the piercing and cutting of high alloy steels, nickel, monel, bronze, copper, brass and aluminium, removal of rivets, and sections of welds. Underwater cutting may be carried out using a hollow steel plastic-coated electrode and specially insulated gun. The arc provides instantaneous heat and the oxygen jet oxidises the iron in ferrous metals.

The *electrode* is held in a special holder and consists of a coated steel tube with a hollow core through which the oxygen passes, the electrode being consumed during cutting.

Equipment

The gun is shown in Fig. 29.1. In addition to the coated tubular electrodes, the equipment comprises a special oxy-arc holder, oxygen cylinder and regulator, a handshield or helmet, and either a d.c. or a.c. welding set up to 300 A.

The *oxygen stream* is controlled by a trigger-operated valve in the holder. During the cutting operation the electrode rests on the plate to be cut, thus keeping the correct distance between oxygen outlet and plate surface.

The high temperature of the arc allows:
1 Cutting to start immediately without pre-heat.
2 Cutting and piercing of oxidation-resistant materials.
3 High speed cutting and piercing of steel.
4 Piercing and cutting of superimposed plates, heavily rusted and corroded parts.

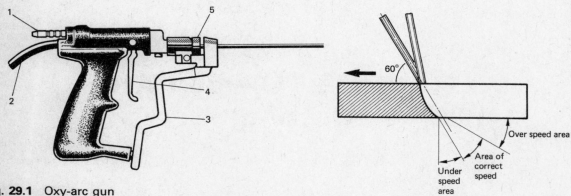

Fig. 29.1 Oxy-arc gun
1 Oxygen Connection
2 Cable
3 Electrode Clamping Lever
4 Oxygen Valve Trigger
5 Oxygen Seal

To Load
Pull back lever (3)
Insert Electrode and push lightly against oxygen seal (5)
Release lever (3)

Fig. 29.2

Cutting Operation

To cut, adjust oxygen pressure and current to the values recommended. Strike the arc, then release the oxygen. Drag the electrode in the direction of the cut with the coating touching the plate. Adjust cutting speed so that the spark stream flows in correct area (see Fig. 29.2).

Advantages

A comparison with oxy-acetylene hand cutting on low carbon steel shows that
1 Where oxygen and acetylene are freely available, the cost per metre is approximately equal for the two processes.

2   The cut with oxy-arc is not so clean and is slightly wider.

3   The speed of cutting with oxy-arc is three to four times faster.

4   Removing rivets and bolts is twice as fast (3 to 7 seconds).

5   Distortion and heat-affected zone are very much less with oxy-arc.

6   Oxygen consumption with oxy-arc is 50–60% of consumption with oxy-acetylene.

7   Oxygen and electric power are consumed only when actual cutting is in progress.

30   The Arc Air Process

Safety

As for other arc processes plus compressed air. Dangers from ejected dross.

Theory and Principles

The process consists of *melting* the metal with a carbon arc and blowing away the molten metal with a parallel jet of high pressure compressed air to form a groove. Ferrous and non-ferrous metals may be cut and gouged, grey cast iron being a difficult metal due to its composition. This process does not require oxidation of the material being cut as in oxy-fuel gas cutting and gouging.

Equipment

This consists of a torch or carbon holder fitted with a dual-purpose cable which carries the direct current (electrode *positive* pole) and a supply of compressed air at approximately 6 to 7 kg/cm^2. The current may be supplied by any d.c. welding generator or transformer rectifier. The electrodes are a combination of carbon and graphite enveloped in a thin layer of copper. The carbon provides resistance to erosion and the graphite provides a lower speed of oxidation and high electrical and thermal conductivity. The copper sheath prevents the electrode tip tapering excessively, allows higher currents, and restricts the radiation of heat as well as generally protecting the carbon/graphite electrode. The torch is shown at Fig. 30.1.

When preparing to gouge, the electrode is inserted in the holder at approximately half its length, with the air jets

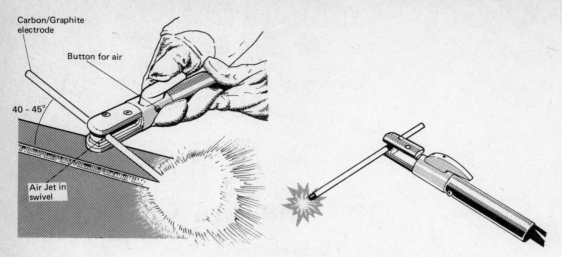

Carbon/Graphite electrode

Button for air

40 – 45°

Air Jet in swivel

Fig. 30.1

Fig. 30.2

positioned *underneath* the electrode to enable the molten metal to be blown away. The gouging angle of the electrode is approximately 45° from the work but may be reduced to 20° for shallow gouging or "washing".

EXAMPLE OF GOUGING SPEEDS

Carbon dia.	Current	Gouge	Speed
6.3 mm	250 amps	6 mm × 10 mm	600 mm/min

Advantages

1 Fast starting and gouging speeds.
2 It is claimed to be faster than all chipping and grinding processes.
3 Surfaces require little or no cleaning, lack of dross.
4 A "cool" process with little fume compared with m.m. arc gouging electrodes.
5 All metals non-ferrous and ferrous may be gouged.
6 Ideal for back gouging of welds and removing defective welds.
7 U-groove preparations for welding may be accomplished using an automatic machine running on tracks and having controlled arc length.

31 Plasma Arc Welding and Cutting

Safety

As for other arc processes plus the following: Danger of severe electric shock from the high open circuit voltage, up to 400 V for cutting. Dangerous fumes and noxious gases when using nitrogen mixtures so it is important to have adequate fume extraction. The intense arc requires a darker shade of filter glass, at least 16 EW (BS 697). Intense high-frequency noise is possible when cutting, especially with non-transferred arcs, of levels 110 dB which requires ear muff protection.

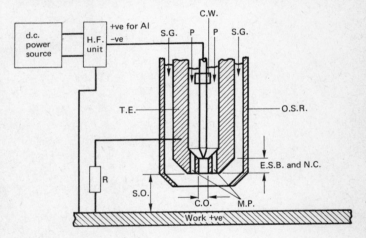

Fig. 31.1 Plasma-arc nozzle set-up
KEY
  C.W. = Cooling water, nozzle and electrode may be water cooled
  P = Plasma gas. Varies with different materials.
  S.G. = Auxiliary shielding gas, usually A+1 to 15% H_2
  T.E. = Tungsten electrode 60°
  O.S.R. = Outer shielding ceramic to prevent double arcing
  R = Resistance limiting pilot arc current (non-transferred)
  E.S.B. = Electrode set back distance
  N.C. = Nozzle constriction
  C.O. = Orifice constriction improves velocity
    2.5 mm dia. 250 amps max
    3.00 mm dia. 350 amps max
  S.O. = Stand-off distance approx. 6 mm
  M.P. = Multi-ports shape the arc plasma and allow increased welding speed
  H.F. = High-frequency discharge ignites the arc

Welding

Theory and Principles

The specially designed copper nozzle has a constricted orifice with a tungsten electrode set back in the body (Fig. 31.1). When an arc is struck, the plasma-forming gas, e.g. argon/hydrogen mixture, is ionised and becomes electrically conductive. This high-temperature ionised gas stream attains a high velocity (near the speed of sound) mainly due to the arc being constricted, and rapidly melts the parent metal. Fig. 31.1 shows the transferred arc method where the arc is established between the tungsten and the work.

Another method used for welding materials which are not electrically conductive is the non-transferred arc method where the arc is established between the tungsten and the copper nozzle. High open circuit voltages (110–120 OCV) and

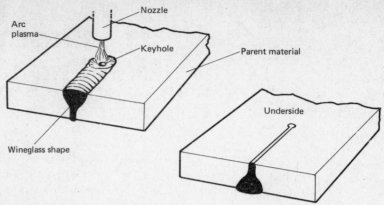

Fig. 31.2 Keyhole method of welding

Nozzle
Arc plasma
Keyhole
Parent material
Wineglass shape
Underside

high-frequency units assist arc starting, or a pilot arc from electrode to nozzle.

The two established methods of welding are (1) the straightforward fusion method, as used when T.A.G.S. welding and (2) the "keyhole" method which uses a higher plasma flow and is usually carried out in the flat position. The force of the high energy arc plasma pushes the weld pool aside to produce a hole (keyhole) and, as the torch moves forwards, the surface tension of the molten metal causes the hole to close behind the torch and form the weld. Fillers may be used with either process. Fig. 31.2 shows the keyhole method, also referred to as the wineglass shape.

Advantages and Applications

1 Most materials may be welded, metal and some non-metals with and without filler.
2 Good penetration control with little distortion, especially on straight butt welds within the range 3–12 mm thick.
3 Good weld appearance.
4 Chemical and aero-engine components. Circumferential pipe welds, etc.
5 High welding speeds up to 10 mm thick with little joint preparation or filler.
6 Narrow H.A.Z. and fewer passes than with T.A.G.S. welding, with faster welding speeds.
7 Micro plasma for thin sheet.

Cutting

Theory and Principles

The transferred plasma arc is used for cutting metals because of the very high energy density and velocity of the plasma jet.

The torch is basically the same as for welding but with

Table 31.1   Materials Welded using Plasma-Arc Welding

Material	Thickness mm	Current/ Voltage	Nozzle dia. mm	Plasma flow & gas l/min	Gas shield	Speed of travel m/h	Comment
Low C steel	3.2	270/26	3.2	2.8 A+H_2	9.5 A+H_2	14	
18/8 Stainless	3.2	145/32	2.8	4.7 A+H_2	16.5 A+H_2	36	
Titanium	3.2	185/21	2.8	3.8 A	21 A	29	
Copper	3.2	60/25	3.2	1.9 A	9.5 A	14	Pre-heat 400°C
Brass 70/30	3.2	180/25	2.8	4.8 A	9.5 A	14	
Nickel	3.2	200/30	2.8	4.8 A+H_2	19 A+H_2	19	

indirect water cooling of nozzle and electrode, stop and start buttons for H.F. start, and multiports around the main arc-constricting orifice. A rectifier-type power source or motor generator is used. A plant of 400 volts OCV of 50 kW at 100% duty cycle is typical.

An H.F. spark initiates an auxiliary arc of low current intensity between the tungsten (cathode) and the nozzle. The plasma gas becomes ionised and electrically conductive and, as the torch is made to approach the workpiece (positive pole), the main arc is automatically struck. The high-temperature high-velocity plasma jet is capable of cutting any metal. Its high velocity blasts away the molten metal in addition to vapourising some to form the cut or kerf.

Cutting Gases

Aluminium and stainless steels require non-oxidising gases for good cutting results in both thin and thick sections. Argon/hydrogen mixtures permit good cuts and high cutting rates because the hydrogen increases the arc voltage and thermal conductivity of the mixture. Parallel kerfs, little dross, oxide-free cut faces and minimal fumes result from the use of A/H_2 mixtures. Argon/Hydrogen/Nitrogen or A/N_2 mixtures are used when machine cutting, but nitrogen is *not* recommended for hand cutting due to the formation of poisonous oxides of nitrogen. Higher cutting speeds are possible with this cheaper mixture with little loss of quality. The increase in cutting efficiency is probably derived from the greater anodic voltage drop associated with the nitrogen gas.

When inert gases such as argon are used, the heat is derived from the electrical energy of the arc. Carbon steels require an oxidising gas for the best results; the exothermic iron-oxygen reaction provides additional heat at the cutting point and so reduces the amount of electric power required. Air has proved to be a most efficient gas.

Cutting Speed

This should be as high as possible for economic reasons provided a narrow kerf and a clean cut at top and bottom edges are produced. For a given electric power and gas mixture, there is an optimum speed range for each type and thickness of material. Excess speed causes a decreased kerf width with an increased bevel but current intensity is the main factor determining kerf width. For manual control and complicated machine cuts 1 m/min is a reasonable speed. In general speeds of several metres/min are used for straight line and trimming cuts.

Table 31.2 Variation of Cutting Speed with Typical Gas-type and Current

Material	Thickness mm	Current amps	Cutting speed mm/min	Gas
Aluminium	1.5	40	1200	A/H_2
	5.0	50	1500	A/H_2
	12.0	400	3750	A/H_2
	25.0	400	1250	A/H_2
Stainless	2	50	1600	A/N_2
steel 18/8	5	100	2000	A/H_2
	12	380	1500	A/H_2
	25	500	625	A/H_2

32 Electron Beam Welding

Safety

X-rays are produced in the welding chamber.

Theory and Principles

Electrons emitted from a heated filament are accelerated by high voltage to a velocity in excess of 160 000 kilometres per second and formed into a long beam only a fraction of a millimetre in diameter. A power intensity is produced more than 5000 times greater than that achieved with a conventional arc.

Fig. 32.1

(*Courtesy Rotax Ltd.*)

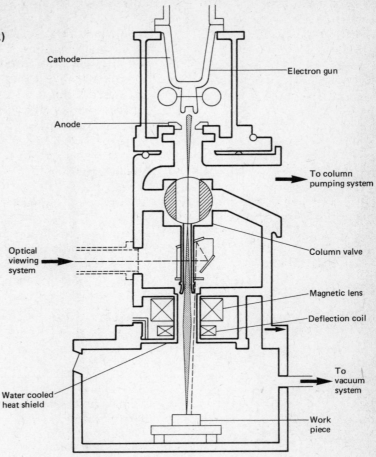

Cathode

Electron gun

Anode

To column pumping system

Optical viewing system

Column valve

Magnetic lens

Deflection coil

To vacuum system

Water cooled heat shield

Work piece

This powerful beam, generated by the electron gun, is precisely controlled and directed down to the workpiece with the aid of a magnifying viewing system that looks down the beam. Impact of the electrons in the workpiece instantaneously melts the material in a very small zone—the tremendous kinetic energy taking the beam right through so that heat is generated directly right across the joint face without dependence on conduction.

Traversing the workpiece with precision servo-operated work-handling equipment, or deflecting the beam as in a television receiver, enables a continuous weld to be produced with consistently high quality, at rates almost up to 40 metres per minute.

Because heat is generated at such a fantastically high rate right across the joint face, only the minimum amount of material necessary for fusion is melted, thus producing maximum joint strength with virtually no distortion or H.A.Z damage.

The equipment is shown in Fig. 32.1

Circumferential E.B. welds

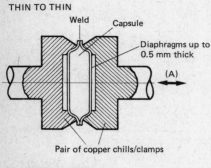

Stainless Steel
to high nickel
for miniature
relays

Fig. 32.2

THIN TO THIN

Weld Capsule

Diaphragms up to
0.5 mm thick

(A)

Pair of copper chills/clamps

(A) May be opened and
closed under vacuum
to ensure vacuum in
capsule for aneroids

Fig. 32.3

Operating Points

Components must be accurately jigged to within 0.5 mm since the minimum spot size is between 0.13 to 0.25 mm for welding thin materials, with a weld zone between 0.6 to 1.3 mm wide. In addition, weld defects such as porosity and lack of fusion are caused through poor fit-up.

Penetration is difficult for refractory metals such as copper and pure aluminium.

Figs 32.2 and 32.3 show two typical applications.

Advantages

1   High joint strengths obtained in normally inaccessible positions.
2   Sound joints in dissimilar metals.
3   Low distortion, high welding speeds.
4   No weld contamination by fluxes, etc.
5   Deep "finger" penetration.

Table 32.1   Typical Joints Welded by E.B.W.

Material	Penetration mm
Magnesium	5.0
Aluminium-magnesium alloy	4.0
Stainless steel	4.0
Nickel	2.3
Low carbon steel	1.8
Titanium	1.5
Aluminium	1.0
Beryllium-copper	1.0
Phosphor bronze	1.0
Copper	0.5
Tantalum	0.5
Molybdenum	0.25
Tungsten	0.25

33   Laser Cutting and Welding

Safety

Laser Radiation May Cause Loss of Sight if Focussed by the Eye Lens on the Retina.

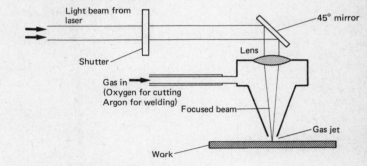

Fig. 33.1

Theory and Principles

Laser cutting and welding depend upon focussing the parallel laser beams by means of a lens so that the energy density is sufficient to melt the workpiece. The carbon dioxide laser system is capable of continuous outputs of 1 kW and 50 kW pulsed.

For welding, argon gas is used. For cutting, oxygen is used in combination with the gas jet. The heat generated at the laser spot varies with different metals and those metals with a high reflectivity and good thermal conductivity (e.g. aluminium, tungsten) are difficult to heat. The stainless steel alloys, although having a bright reflective surface, heat readily because of their relatively poor conductivity.

To cut materials, the high power output produces a more powerful beam which, when combined with the oxygen jet, causes the metal to boil and then vapourise. The fluid oxides of the metal formed are then expelled from the kerf by the action of the jet. The width of the kerf is determined by the size of the focussed laser spot.

Advantages

1   Little buckling and distortion.
2   Virtually any material may be cut.
3   Very narrow kerf width (0.4 mm and less).
4   High cutting accuracy, especially when automatic profile cutting.

Table 33.1 Some Materials Cut by Laser Gas-jet Techniques at a Power of 200 W

Material	Thickness mm	Gas	Speed mm/min
Low carbon steel	0.5	Oxygen	635
Stainless steel	0.5	Oxygen	2600
Titanium	0.6	Air	200
Zirconium	0.25	Air	915
Carborundum (sintered)	1.6	Air	760
Asbestos cement	6.3	Air	25
Glass (soda-lime-silica)	4.0	Air	100
	1.6	Air	380
	0.2	Air	5000
Perspex	25.0	Air	100
	10.0	Air	200
	4.6	Air	635
Nylon	0.8	Air	5000
P.T.F.E.	0.8	Air	6100
G.R.P.	2.4	Air	635
Leather	3.2	Air	635
Wood: Deal	50.0	Air	100
Oak	18.0	Air	200
Teak	25.0	Air	75

34 The Technique of Soft Soldering

The soldering bit shown in Fig. 34.1 is one form of several to suit different applications, all of which work on the same principle. Other types are electrical or gas heated with straight (hatchet) or pencil-shaped bits. The high conductivity copper bit acts as a reservoir for heat and molten solder and also as a means for delivering the molten solder to the job.

Before soldering starts, the bit must be "tinned". A suggested method is as follows.

1 Heat the copper in the gas stove until it melts the solder freely (remove when flame turns green).

2 Clamp in a vice and file the surfaces of the point until they are clean and smooth. Round the point slightly as well as the sharp edges.

3 Re-heat and apply flux to the faces of the bit and add a little solder until all sides are tinned.

Fig. 34.1 Copper soldering bit

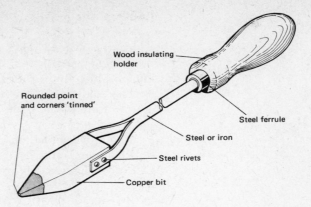

Wood insulating
holder

Rounded point
and corners 'tinned'

Steel ferrule

Steel or iron

Steel rivets

Copper bit

Pre-tinning of Laps

1 Clean lap area with emery cloth and brush on flux sparingly.
2 Apply flat face of bit to one end of lap and apply solder to bit end.
3 Re-heat and apply flux to the faces of the bit and add a little solder until all sides are tinned.
4 Clean off any excess flux with a wet rag.

Methods of Soldering (Bit Soldering)

1 Remove dirt, grease or oxide from surfaces to be soldered if necessary, and apply a thin coating of flux.
2 Place the pieces to be soldered on a support, but not a large piece of metal which will absorb and take away the heat. If necessary, clamp the parts together or use U-shaped spring clips.
3 Take the heated tinned bit and apply solder to the end. Apply the bit end on the seam until the heat penetrates and the solder is drawn and flows between the two surfaces.
4 Move the flat face of the bit along the seam and hold firm with a piece of wood until cool. Long laps may be tacked or pre-tinned and "sweated" together by moving the heated bit slowly along the lap.

Flame Soldering

The flames are usually natural-gas air or propane air. For pewter or Britannia metal the flame must be a small one and the heating time must be extremely short. Larger blowpipes are used for filling in dents on vehicles or radiators or capillary fittings. Automatic flame soldering is used in conjunction with rotating conveyors in factories for soldering boxes and cans, with the solder preplaced.

Applications

1   All pre-formed joints such as cans and food containers.
2   Sealing paned down or internal grooved joints to make liquid tight containers.
3   Lap soldering of tabs and handles.
4   Electrical work, such as connections between cables, bars, tubes and nipples.
5   Radio and television circuits and fuses, etc.
6   Plumbing work.
List other applications you are familiar with.

Welding Demonstration 9 and Investigation 17 Soldering Fluxes and the Metal Surface

AIM   To demonstrate the use of different fluxes on different types of metal surfaces.

THEORY   Certain elements in metals, when in contact with oxygen in the atmosphere, form oxides, which prevent the tin in the tin/lead soft solders from surface alloying with the metals to be joined. The fluxes used must be capable of removing this oxide and the more tenacious (difficult to remove) the oxide, the more powerful (active) the flux must be. An example is the highly active concentrated hydrochloric acid on chromium oxide in stainless steel.

EQUIPMENT AND MATERIALS   Tongs, gas stove, electric iron, natural-gas air torch, copper bit, F and K type solder (See Table 25.1).

Fluxes: zinc chloride, resin type (easi-flow type), zinc ammonium chloride (paste or Baker's Fluid No. 1).

Hydrochloric acid (50% conc.). Asbestos support, small G-clamp and flux brush and wood.

Metal samples: 80 mm × 40 mm—two off each, copper, low carbon steel, lead/tin coated steel (Ternplate), tinplate, galvanized steel, 18/8 type stainless steel and two pieces of copper cable.

PREPARATION   All laps cleaned thoroughly and shear burrs removed.

PROCEDURE

Solder the metal samples using a 15 mm lap with correct clearance and flux and solder recommended (see Results table). The bit and flame may be used to demonstrate, and pre-tinning of the surfaces may be used to demonstrate this advantage especially with the stainless steel and low carbon steel. Use the electric iron to solder the cable ends. The joints to be cut in order to note capillary flow and peel tested. Other combinations of heat, flux and solder may be tried. Now complete the Table.

PEEL TEST   Place specimen in vice and attempt to peel one lap from the other by means of pliers.

SAFETY
1)   Avoid touching very hot metal with bare hands.
2)   Beware of acid flux splashing on hands, face and clothes, and do not rub eyes during the soldering operation.
3)   Great care must be exercised in the use of concentrated hydrochloric acid as severe burns can result from contact with the skin.

RESULTS

Metal	Flux	Grade of Solder	*Visual Inspection	*Peel Test	Remarks
COPPER	Zinc chloride or Resin	F			
LOW CARBON STEEL	Zinc ammonium chloride	K			
TERN PLATE	Resin or Zinc chloride	F			
GALVANIZED STEEL	Zinc ammonium chloride	K–F			
18/8 STAINLESS	50% Hydrochloric acid (*Care*)	K			
COPPER CABLE	Resin	K			

* Mark: Excellent, Satisfactory, Poor.
Note: Ensure that all sheared edges on coated sheet are soldered.

QUESTIONS
1)   What is best quality Tinman's solder composed of?
2)   What is Ordinary grade solder composed, approximately, of?
3)   What is Plumbers solder composed, approximately, of?
4)   Which element causes solder to become more fluid and have good electrical conductivity?
5)   What does hydrochloric acid dissolve on the metal surface?
6)   Which flux is the non-active flux and which the active flux?
7)   Why is it better to tin low carbon steel before soldering?
8)   What is the term used when the solder is drawn into the laps?
9)   Place a piece of solder between two fluxed copper laps and see which is the quickest method of soldering—flame, copper gas heated bit, or electrically heated bit.
10)   Would you say a soft soldered stainless steel joint is used as a seal or a strength joint?

11) The approximate highest melting point for tin/lead soft solder is
a) 500°C *b*) 300°C *c*) 100°C *d*) 50°C
12) The strongest joint is formed when there is a thin/thick layer of solder between the faces.
13) Good quality Tinman's solder melts at
a) 183°C *b*) 223°C *c*) 53°C *d*) 103°C

35 Joining Pipes and Conduit by Mechanical and Thermal Methods

Type of Joint and Surface Preparation

The methods used for joining pipes fall into two distinct categories:
1) Mechanical, and
2) Thermal

The mechanical methods, such as screwed sleeves and flanges, bolting, cold riveting and flare-type taper unions, are generally not designed to withstand such severe working conditions as the thermal joints (with exceptions noted below).

Screwed Joints
This method of joining uses either a screwed sleeve which may be straight, or an elbow, for turning the pipe or conduit through ninety degrees. Cross-shaped junctions are used for joining four branches also. The junctions may be malleable iron or forged steel, depending on the service requirements.

When gas or water is to be conveyed, some form of seal is required between the pipe threads, such as hemp and litharge or the various proprietary compounds.

Note the jointing methods for the heating system in your College.

Flanged Joints
The flange may be attached to the pipe either by screwing or by welding but it is important in both cases that the two mating flange faces are parallel and correctly prepared. Invariably there is some form of seal between the faces to seal the joint adequately when the bolts are tight around the flange. The type of seal is dependent on the pipe contents, i.e. steam, oil, acid, water, etc., and may be of asbestos, fibre, butyl or

proprietary packings of stainless steel and fibre such as "Metaflex".

Often the flange faces are prepared with a gramophone spiral on the face.

Riveting of large diameter pipes has now been superseded by welding, but repairs must still be carried out.

Welded Joints

When pipes have to carry steam, water, oil or gas under pressure, the specification usually requires that the joints be fusion welded. If the joint is not subject to any great stress, it may be brazed or soldered to form a water- or gas-tight seam. Fusion welding of steel pipes by the oxy-acetylene process is widely used in the heating and ventilating industry, and for small-bore pipes in boilers. The branch pipes are usually scarfed by the oxy-acetylene cutter and dressed by a file or grinder, then the hole in the body is cut out, and then the branch "set in" (Fig. 56.1B, p. 194) to ensure satisfactory penetration and fusion. On low pressure work, the branch may be "set in" (see Fig. 56.1A). When pipes carrying high temperature and high pressure steam are welded, such as are found in power stations, close control and fit-up is essential to ensure top quality welds. Often backing rings and E.B. inserts are used to provide uniform penetration and location. When non-ferrous pipes or low alloy steel pipes are welded, gas shielded root runs are deposited and then capped by manual metal arc.

Soft and hard soldering of pipes and conduit is employed where brass or copper pipes require joining.

36 Applications Involving Pipe

1 JOGGLED AND RIVETED

Application Used on hydro-electric pipes for water under pressure.

Location On the left of the "Miner's Track" going up Mount Snowdon (Wales).

2 OXY-ACETYLENE WELDED (FUSION)

Application Joining small-bore steel pipe carrying gas under pressure or hot water heating systems.

Location In the College welding workshop, and from the hot water boiler.

3 METAL ARC OR M.A.G.S. WELDING (FUSION)

Application Large bore pipe butt welded for gas

Location Natural gas pipeline from Canvey Island in the South (England), to the North, with branches to East and West. Denoted by a bright red sign.

4 METAL ARC WELDING (FUSION)

Application Large bore thick wall (50 mm) low alloy, high pressure steam pipe: Vee or U preparation.

Location Nuclear Power Stations such as Trawsfynydd (Wales), Calder Hall (England), etc.

5 SILVER SOLDERING (NON-FUSION)

Application Copper and stainless steel pipes for hot water and hot gases.

Location Rolls Royce Engines and water evaporators.

INVESTIGATION

List as many methods of joining pipe and conduit as you can find. Make a sketch of the joint and type of joining method, and say what the content of the pipe is.

37 Material Removal

Basically, there are two methods of removing metal mechanically:

a) Chip forming, i.e. drilling, planing, end milling, sawing.

b) Non-chip forming, i.e. tin snips, shears, punches, croppers, notchers, nibblers and guillotines, all of which employ a form of shearing action.

Shearing (Guillotine)

The principle of shearing is similar to punching except that the area being sheared is a relatively small continuous section, starting at one end of the plate and ending at the other. A hold-down clamping stop holds the plate rigid during cutting in the case of guillotines. (See Fig. 37.1.)

An important factor in the production of a good cut edge is the clearance between the blades as well as the sharpness of the blade edges. Some machines have provision for altering the blade clearance to suit the thickness of plate being cut, but usually the clearance is set at manufacture and checked periodically. A reasonable guide is 0.1 mm increase in clearance for every 1 mm of sheet thickness.

Fig. 37.1 Guillotine: principle of shearing

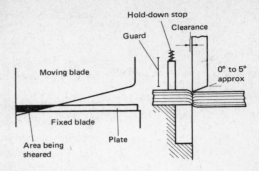

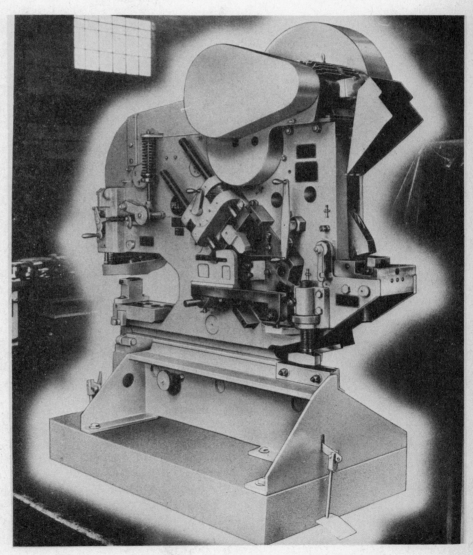

A Universal Machine incorporating all the necessary mechanical cutting facilities for fabrication. Note the fixed hinged guard top right for the notcher. (*Courtesy Henry Pels Ltd.*)

Shears

There are different kinds of shears, both for thin and thick sheet. The following are the most common types, in addition to the guillotine.

1 Rotary Shears

a) Usually mechanically powered but may be hand operated for thin sheet.

b) May have one straight fixed blade and one rotating circular blade (Fig. 37.2).

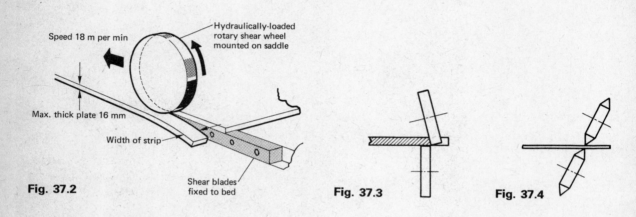

Speed 18 m per min

Hydraulically-loaded rotary shear wheel mounted on saddle

Max. thick plate 16 mm

Width of strip

Shear blades fixed to bed

Fig. 37.2

Fig. 37.3

Fig. 37.4

c) May have two rotating circular blades (Fig. 37.3), one at an angle to the other.

d) Two rotating circular bevelled blades set at an angle to the sheet, but in line (Fig. 37.4).

2 Hand Lever Shears (or Knife)

May have a straight or curved top blade. The force is obtained from leverage on a long handle.

3 Plate and Flat Bar Shears (Mechanical power)

With one fixed and one moving blade, similar to a guillotine, but faster and with a shorter blade (Fig. 37.5).

4 *a)* Cropper

For shearing angle section and other sections. Usually has one fixed and one sliding blade to take the section shape (Fig. 37.6).

b) Notcher

For mitres and removing square and rectangular shapes. Often there is a combination of punch, cropper, notcher and shear named a Universal Machine.

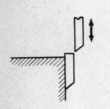

Fig. 37.5

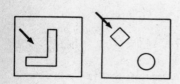

Fig. 37.6

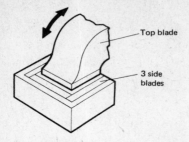

Fig. 37.7 Notcher

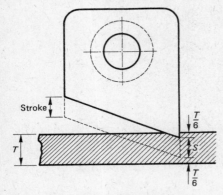

Fig. 37.8 Nibbler shearing
$S = \frac{2}{3}T$ and $T = 3$ mm max

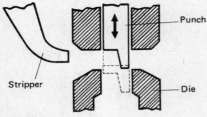

Fig. 37.9

Tin Snips

There are numerous types, but all work on the same principle of leverage combined with hardened blades ground to a cutting angle. The nearer the metal edge being cut is to the fulcrum of the shear blades, the greater the force exerted. This is, in turn, governed by the metal thickness; hence the reason why comparatively thin metals only are sheared by this method. The handle length which provides the leverage on the blades varies to suit the sheet metal thickness being cut, but again, as one hand only is used to exert the effort, thick metals push the handles open too far.

Make a sketch of a typical pair of tin snips and show:
a) the length of leverage; *b*) the effort;
c) shearing load; *d*) fulcrum.

Nibbler Shearing

See Fig. 37.8. This is a high-speed shear for thin sheet.

Nibblers

These vary in capacity and can deal with plates and sheet from gauge plate up to approximately 10 mm thick. The action is a reciprocating punching action, utilising a special top tool and a bottom die and also a stripper (Fig. 37.9). Circular slots may be nibbled by using a fixed centre on the plate and revolving the plate during nibbling.

Portable nibblers for up to approximately 3.5 mm sheet use a special round hollow punch working with the aid of a carrier pin at which the die is hung centrally. The carrier pin absorbs the punching pressure when nibbling and ensures that the punch and die are concentric during nibbling. The stripper often serves as the punch guide and may be adjusted in height.

38 Principles of Punching

Punching is a method of rapidly removing metal to form round, square, elliptical or other shaped holes by using a top punch and bottom die. The punch is usually driven by mechanical, hydraulic or manual force, dependent on the thickness of metal. A line diagram of the punching arrangement is shown in Fig. 38.1.

Cropper for cutting mitres and straight cuts on angle section. Note the setting angle for various mitre cuts and the hold-down stop in position on the angle. The square and round hole right and below are for cutting bar stock. (*Courtesy Henry Pels Ltd.*)

Shearing arrangement for straight cuts for flat bar and angle legs. Note the position of the hold-down stop immediately above the section being cut. (*Courtesy Henry Pels Ltd.*)

Notching arrangement for square and mitre cutting of flat section and angle legs. Note how the blades may be interchanged, and the backstop screw which may be re-moved or adjusted. (*Courtesy Henry Pels Ltd.*)

Cropper with special blades for cutting channel sections. Note the removal of the angle cropper arrangement on left. (*Courtesy Henry Pels Ltd.*)

Punching arrangement. Note, bottom left, the moveable bolster and die. The punch directly above is circled by the plate stripper. The lever top right actuates the punch. (*Courtesy Henry Pels Ltd.*)

Fig. 38.1

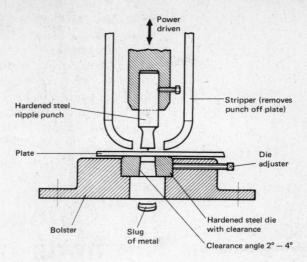

The action of the punch in forcing the slug of metal through the die is, first, plastic flow of the material due to the compressive load, then one of shearing and tearing as the grains of the metal are elongated and ruptured. This results in severe work-hardening and an increase of stress around the hole. Specifications for certain types of work forbid punching and others only allow it if the holes have been reamed through to remove this highly stressed work-hardened metal. B.S. 449 does not recommend the punching of holes in plates above 15 mm thick and as a rule it is not possible to punch holes smaller than the plate thickness without damage to the punch.

A clearance between the punch and die of approximately one-tenth of the plate thickness is usual to prevent excessive burr and roll over.

Safety

Before using any power-operated machine for material removal such as punching, shearing, notching or nibbling, etc., it is essential that the following is understood:

1 Do not use any machine unless you have had operating instructions from a responsible person.

2 You should know the position and operation of emergency stop buttons or switches.

3 Before starting a machine, ensure that no one is at the back in a dangerous position, and that all guards are in place.

4 Keep fingers clear of blades, hold down stops and stripper (see Fig. 38.1).

5 Ensure that the work is held down tight during shearing.

6 When handling plates having a sharp burr, wear leather hand protectors.

INVESTIGATION

Obtain a piece of low carbon steel of sufficient thickness to be able to be punched on your College punch. Punch a hole and then drill one next to it the same size, cut through the holes and file and emery to a fine finish.

Etch the edges and inspect under a magnification sufficient to show the grain structure.

Sketch the results ×4 full size to show the difference.

39 Comparison of Methods of Preparation of Edges and Surfaces

It is most important that any surface which will have contact with a molten filler is clean and free from heavy oxides, scale, grease and paint.

Butt, corner and fillet joints invariably need some form of preparation of the plate edges. The choice of the method of preparation depends upon several factors, the more important being:

1 Type of edge preparation specified.
2 Type of material and thickness.
3 Equipment and machinery available, including welding process.
4 Accuracy of cut required (depends upon class of fabrication).
5 Length of edge to be prepared and number off.
6 Whether the prepared edge is straight or curved.
7 Thickness of material.
8 Total time of preparation (setting up—preparation—cleaning).
9 Cost.
10 Class of work.
11 Joint location and configuration.

Welding Investigation 18
Preparation of Single Bevel Butts and Straight Butts

EQUIPMENT Quantity of 150 mm × 100 mm × 6.5 mm plates. Various edge preparation equipment. Safety goggles and equipment.

1)   Set up and prepare the plates and note the relative times. Then draw and complete the Table.

2)   Mark two plates down the centre lengthways and cut one on the guillotine and one with the oxy-fuel gas hand cutter and complete the Table.

SINGLE BEVEL BUTT

Method of Preparation	Total Setting-up Time	Actual Preparation Time	Order of Accuracy 1 to 4	Quality of Cut Face	Remarks (versatility, cost, number off)
Hand hammer and chisel					
Pneumatic chisel					
File					
Hand oxy-flame cut					
M/c oxy-flame cut					
Pneumatic grinder					
Bevel shear					
Electric grinder					
Milling machine					
Plate edge planer (mainly for comparison)	4 m length. 12.5 mm thick 30 min.	20 min.	1	Excellent	Can produce V, U or straight butt stack of plates planed

STRAIGHT BUTT

Method of Preparation					
Hand oxy-flame cut					
Guillotine					

Single Hem

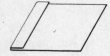

Double Hem

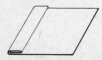

Fig. 40.1

40   Stiffening of Fabricated Material

Folding Edges

The width of the hem W (Fig. 40.1) may vary between 5 and 10 mm, depending upon type and thickness of material and application, and allowance must be made for metal thickness.

Jay preparations being milled on thick plate using an edge milling machine. Note the hydraulic clamps on the left, the protractor for setting the edge bevels. (*Courtesy Hugh Smith (Glasgow) Ltd.*)

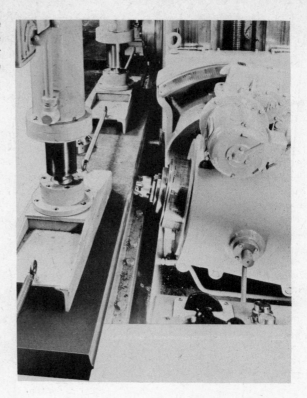

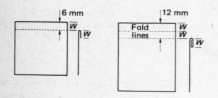

Fig. 40.2

Laying Out
Examples are given in Fig. 40.2.

Example Calculation
For double hem: $W = 6$ mm
Allowance $= 2 \times W = 12$ mm

Forming Analysis   (Using a bar or a cramp folder)
1   Set folder for the correct width of the hem (allow for metal thickness on cramp folder).
2   Insert sheet and swing bending beam as far as it will go.
3   Remove sheet and place it with the bent edge facing upward on the bevelled part of the blade. Flatten the hem by bringing the beam down, or by malleting.
4   To form a Double Hem, repeat the above procedure.

Comparative Uses
The hem is used to eliminate sharp edges, increase strength and stiffness, and improve appearance. The hem is the simplest and quickest method of forming a safe edge.
   Uses: Bins, trays, table tops, chutes, boxes and shelves.

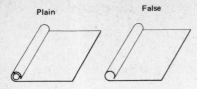

Plain   False

Fig. 40.3

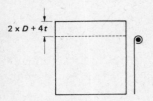

$2 \times D + 4t$

Fig. 40.4

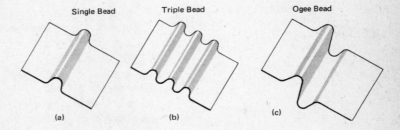

Fig. 40.5

Wired Edges

The two types are shown in Fig. 40.3. Wire allowance is shown in Fig. 40.4.

Example Calculation
$$D = 3 \text{ mm}, \ t = 1 \text{ mm}$$
$$\text{Wire allowance} = 2 \times D + 4t$$
$$= 6 \text{ mm} + 4 \text{ mm} = 10 \text{ mm}$$

Laying Out (Plain and False)
$D =$ dia. of wire
$t =$ thickness of sheet

Forming Analysis
Wired edges may be formed by manual methods or on a wire burring machine. Cylindrical objects may be wired first and the cylinder formed later.
1   The allowance is first marked on the edge of the sheet and folded to approximately 60–70°, or bent over the edge of a rounded block (Fig. 40.5).
2   The wire is now placed in the fold and the metal folded round the wire by striking with a mallet on a flat surface.
3   The edge is now closed either by striking with square-faced hammer, taking care not to mark the metal, or fed through the wheels of a burring machine.

Comparative Uses
Similar to hemmed edges, but stronger and stiffer.

False Wired
Used on lockers, shelves, boxes, edging on canopies, hinges, light gauge angles, waste bins, draining boards, tea-urns.

Swaging

Examples are shown in Fig. 40.6.

Fig. 40.6

Single Bead   Triple Bead   Ogee Bead

(a)   (b)   (c)

Laying Out
The size of roller to be used and the number and type of beads must be ascertained, and the extra metal allowed for on the

pattern. A piece of scrap sheet may be used to find the allowance.

Forming

1 The sheet is marked off where swaging (beading) is required.
2 The sheet is fed between the swaging rolls, gradually increasing the adjusting screw pressure until the depth of swage is reached.

Comparative Uses

Used to stiffen and strengthen flat panels and cylindrical containers and to remove slackness in large areas of flat plate or to give an ornamental finish.

Swaging Examples

The shape of the bead depends on the wheels. A typical set of swage wheels is shown in Fig. 40.7.

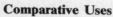

Fig. 40.7

41 Self-secured Joints in Light Gauge Fabrication

Paned-down

The joint is illustrated in Fig. 41.1 Laying-out examples are given in Figs. 41.2, 41.3, 41.4.

Example Calculation
    Mean diameter $D = 440$ mm
Allowance width $W = 5$ mm
        Height $H = 200$ mm
Round Bottom:
    Overall dia. of bottom $= D + 4W$
$$= 140 + 20 \text{ mm}$$
$$= 160 \text{ mm or } 80 \text{ mm radius}$$

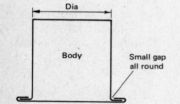

Fig. 41.1 Paned-down joint
Note: The small gap (approximately 1 mm) is left to prevent distortion during forming

Comparative Uses

This joint is the simplest form of self-secured joint and may be used for bottoms on circular, square and rectangular work or for joining a cone to a cylinder, or blanking off the ends of ductwork.

It is not a particularly strong joint and should be avoided if there is the possibility of rough usage. It is seldom used on material thicker than 1.5 mm.

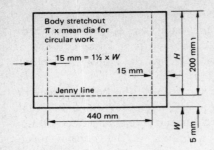

Fig. 41.2

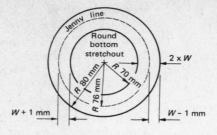

Fig. 41.3 Example of round-bottom layout

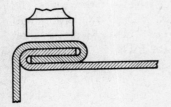

Fig. 41.4 Example of square-bottom layout

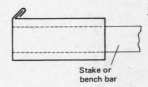

Fig. 41.5 Knocked-up seam   **Fig. 41.6**

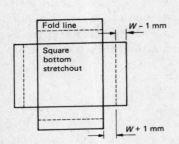

Fig. 41.7

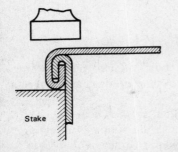

Fig. 41.8

Knocked-up Seam

The joint is given in Fig. 41.5. Pattern layout and calculation as for paned-down joint.

Forming Analysis

1   After forming the paned-down joint, place the container over a suitable stake with the inside edge firmly against the end of the stake and with a mallet, bend the seam over gradually to approximately 45° (Fig. 41.6).

Form the bend by gradually turning the container constantly.

2   When the seam has been completely flattened, tighten the joint with a light hammer (Fig. 41.7), remove the container from the stake or bar, and rest the edge on a flat stake. Now tap the bottom lightly to square and straighten the seam (Fig. 41.8).

Comparative Uses

This joint is much stronger than the paned-down joint and is also used for fitting bottoms on to round, square or rectangular containers, when access may be gained to support the knocking-up operation. It is the most common self-secured joint.

Circular body stretchout $= \pi D +$ joint allowance (see grooved joint)

$$= \tfrac{22}{7} \times \tfrac{140}{1} + \text{J.A.}$$
$$= 440\,\text{mm} + \text{J.A.}$$

Body height $= H + W$
$$= 200 + 5\,\text{mm} = 205\,\text{mm}$$

161

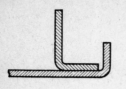

Fig. 41.9

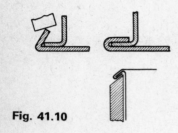

Fig. 41.10

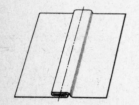

Fig. 41.11 Grooved seam

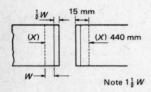

Fig. 41.12 Layout of grooved seam x shows the calculated length of cylinder before rolling

Fig. 41.13

Fig. 41.14

162

Forming Analysis

1 Mark out one parallel line from the edge of the body stretchout (Fig. 41.2) and two parallel lines from the edge of the round bottom to dimensions shown in Fig. 41.3, and cut out.

2 Form a flange of the correct width on the bottom piece on the jenny, keeping the edge firm against the guide. Next, jenny the edge up on the body using the same guide setting.

3 Insert the body of the container in the bottom piece, making sure it is a loose fit, as in Fig. 41.9. Then using a paning-down hammer, after pinching the container at four diametrically opposite places, close down the flange gradually to prevent buckles and wrinkles forming. Ensure that the work is on a flat surface and turned constantly. The final paning down is done using the peen end of the hammer, or by inverting over a hatchet stake and closing (Fig. 41.10). Care should be taken not to mark the body.

Grooved Seam (Inside or Outside)

A grooved seam is shown in Fig. 41.11. Laying-out in Fig. 41.12.

Allowance

The allowance for folding and grooving is three times the width of the desired groove, and is marked one and a half times on each edge as shown. For metals heavier than 26 s.w.g., allowance must be made for folds.

Note: The groove width varies to suit the job.

Example Calculation
Mean diameter $D = 140$ mm
Groove width $W = 10$ mm
Calculated Circumference (x)
$$x = \pi \times D = \tfrac{22}{7} \times 140 = 440 \text{ mm}$$
Joint Allowance (J.A.)
$$\text{J.A.} = 1\tfrac{1}{2} \times W = \tfrac{3}{2} \times 10 \text{ mm}$$
$$= 15 \text{ mm on each edge}$$

Forming Analysis

1 On the full stretchout length, mark off two parallel lines from the edges at $\tfrac{1}{2}W$ and $1\tfrac{1}{2}W$ from the edge, as shown in Fig. 41.12.

2 Bend the two edges in the folder as shown in Fig. 41.13 and then roll, taking care not to close the bends by inserting a thin shim. Now hook the bend edges together (Fig. 41.14).

3 Select a suitable bench bar, stake or mandrel and a suitable grooving tool which should be slightly larger than the desired width of fold, 1 to 2 mm depending on the metal thickness.

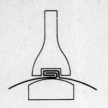

Fig. 41.15

4 Fit the groover over one end of the seam at a slight angle to begin with and strike the groover with a hammer. Repeat this at the other end of the seam to prevent the edges coming apart. Then groove the entire seam in stages (Fig. 41.15) and finally tighten the seam by flattening with a square-faced hammer.

Comparative Uses

The grooved seam is used to join two pieces of flat sheet metal or to fasten the sides of round or square articles such as trunking, ducts and containers and vessels. The seam may be made by hand or machine.

42 Folding Simple Components

Figure 42.1 shows an example of folding and forming in correct sequence. Try it yourself in tinplate or paper and glue, and then try other sequences.

Manufacture the example in Fig. 42.2 in tinplate and solder the seams.

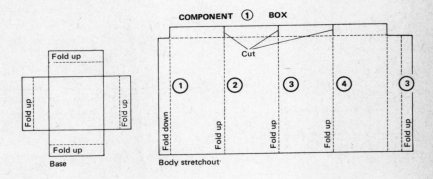

Fig. 42.1

(continued on page 164)

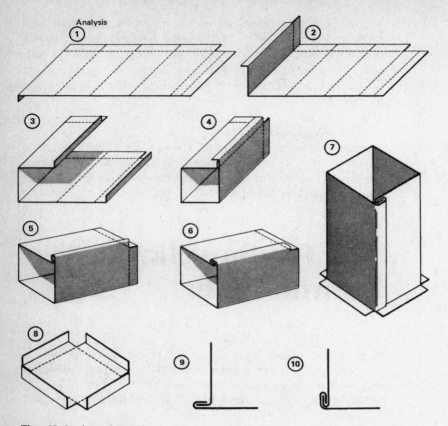

Analysis

Fig. 42.1 (continued)

Fig. 42.2

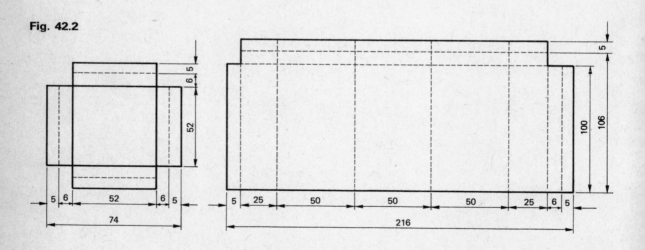

43 Tinman's Heads and Stakes

1 Bick iron
2 Anvil stake
3 Creasing iron
4 Combined funnel and side stake
5 Grooving stake
6 Horse
7 Saucepan belly stake
8 Round bottom or canister stake
9 Tinman's mandrel
10 Hatchet stake
11 Funnel stake
12 Side stake
13 Extinguisher stake
14 Half-moon stake
15 Pipe stake
16 Blow horn stake
17 Oval horse head

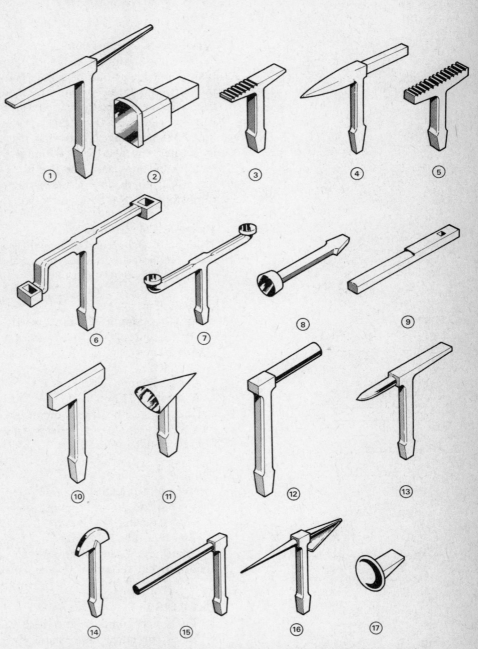

A selection of the more commonly used heads and stakes. Stakes are manufactured from steel forgings, heads from cast steel or cast iron.

44 The Neutral Line

Theory

When metal is rolled into a circle, curved or bent by flanging or pressing, the outer radius of the material is always greater than the inside radius. This results in the fibres of the metal on the outside bend becoming stretched in tension, and the inside fibres becoming crushed in compression. Approximately midway between the two (this varies with very soft and hard metals) there is a line known as the *neutral* or *mean line* which is neither stretched nor compressed and remains the same length. It is important that this line is worked to when calculating metal length and allowance before rolling and bending.

Examples of the neutral line are given in Figs. 44.1 and 44.2.

Example Calculations

Take π as 3.142 and material as low carbon steel (normalised). (It should be noted that other methods are used in conjunction with B.A. tables.)

Bracket (Fig. 44.1)
To produce length of flat required:
$t = 6$ mm
$A = 20$ mm
$B = 20$ mm
$C = 23.56$ mm (calculated)
$R = 2 \times t = 12$ mm
Calculated length $C = \pi \times$ mean dia. $\div 4$ (quarter circle)
$$= (3.142 \times 30) \div 4 = 94.26 \div 4 = 23.56 \text{ mm}$$
Length of flat required $= A + B + C$
$$= 20 + 20 + 23.56 = 63.6 \text{ mm.}$$

Cylinder (Fig. 44.2)
$t =$ metal thickness $= 3$ mm
$A =$ outside dia. $= 116$ mm
$B =$ inside dia. $= 110$ mm
$C =$ mean dia. $= 113$ mm
(measured to neutral line).
Length of flat required $= \pi \times$ mean dia.
$$= 3.142 \times 113 = 355 \text{ mm.}$$

EXERCISES
1   Draw and complete the table (page 167).
2   Try a practical experiment by calculating the length required for a given cylinder diameter using flat bar. Bear in mind the minimum diameter and maximum thickness your workshop rolls will manage.

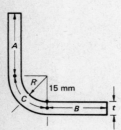

Fig. 44.1

15 mm

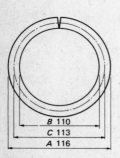

B 110
C 113
A 116

Fig. 44.2

Required Flat Length (mm)	Outside diameter (mm)	Inside diameter (mm)	Metal thickness (mm)	Mean diameter (mm)	Practical experiment
	200	188			
	116	110			
	900		10		
	612			600	

45 Bending and Bending Allowances

Fig. 45.1 Folding machine

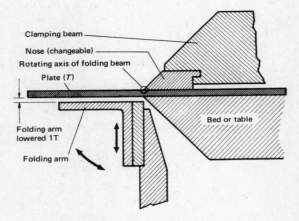

Clamping beam
Nose (changeable)
Rotating axis of folding beam
Plate (*T*)
Bed or table
Folding arm lowered 1T
Folding arm

Sheet Material

When sheet material is to be bent in a folder of the type shown in Fig. 45.1 several important points must be noted, i.e.

1   Any sharp edges such as shear burr should be removed off the fold line, otherwise small cracks will appear on sharp radii.

2   The folding arm must be lowered below the bed for one thickness of the metal being bent, plus the thickness of any nose radii if used.

3   An allowance must be made from the line of sight (directly above the nose of the clamping beam) to the bend line (Fig. 45.2). The bend line is at the centre of the metal forming the bend.

4   Once bend allowances have been worked out in practice for a particular machine, they should be noted and adhered to. There are numerous types of machine for bending sheet metal and plate.

5   A formula derived from practical workshop use is:

Bending Allowance (B.A.)
$$= (0.01743 \times R + 0.0078 \times T) \times L$$

where R = internal bend radius
T = thickness of sheet
L = degrees through which metal is bent.

6   If the bending allowance on folded joints is insufficient this results in internal and external dimensions being too small, and if the bending allowance is excessive, then the internal and external dimensions will be too large. Allowance must be made for material thickness also.

Fabrication Investigation 1
Bending Allowance

AIM   To make a table of bending allowances by calculation and practical experiments so that equal legs of angle are obtained at 90° in low carbon steel sheet.
EQUIPMENT   Folder. Various thicknesses of metal, all the same length and width, i.e. 1 mm, 1.6 mm, 3.2 mm, 4.8 mm. Rule, square, scriber, marker (a vernier height gauge may be used for greater accuracy).

PROCEDURE
1)   Check T with micrometer, R with radius gauge, then mark a line in the centre of each piece of metal and stamp the end which is to go under the clamping beam. Now mark the line for the bending allowance (B.A.) (see Fig. 45.2) obtained by calculation.
2)   Lower the folding beam from edge of table an amount equal to metal thickness (Fig. 45.3). Insert the metal and clamp, then bend to 90°, but allow for "springback", then check each leg for being square and equal. When the correct bend is arrived at, note in the Table the B.A. used. (Try a piece of scrap plate to ascertain the inside radius first.)

Line of sight

Bend allowance

Fig. 45.2

Neutral line rad = 0.111

$R = 0.1$

$T = 0.022$

Allowance 0.17

Fig. 45.3

EXAMPLE CALCULATIONS FOR B.A.

1   *Using Workshop Formula:* $R = 0.1$, $T = 0.022$
B.A. $= (0.017\,43R + 0.0078T) \times 90$
$= (0.001\,743 + 0.000\,1716) \times 90$
$= 0.001\,9146 \times 90 = 0.172\,314$ (say 0.172)

2   *Using Neutral Line Method:*
B.A. $= \pi \times D \div 4$
$= 3.142 \times 0.222 \div 4$
$= 6.975\,24 \div 4 = 0.1743$ (say 0.174)

Sheet thickness (mm)	Inside rad. (mm)	B.A. (mm)	Plate length (mm)	Leg 1 (mm)	Leg 2 (mm)	Wrong	Correct
1							
1.6							
3.2							
4.8							

46   Pipe and Conduit Bending

When bending pipe and conduit, care must be taken to avoid excessive stretching of the outer radius, which causes thinning of the wall thickness and flattening. Corrugating of the inside radius must also be avoided.

To avoid the above defects, circular grooved formers are used of a set standard radius. Levers in the case of conduit and hydraulic rams in the case of heavier pipe are used to form the bend. The pipe is often filled with sand, low melting point alloy or a bending spring or mandrel. In the case of heavy walled pipe, local heating is used to assist in forming the bend with the aid of a ram.

Note: No bending must be carried out in the blue brittle range of temperature, i.e. 200 to 500°C.

The bending allowance is as shown in Fig. 46.1 and the minimum radius used is usually three times the diameter of the pipe from the centre line.

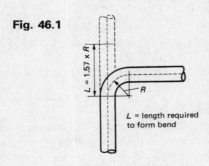

Fig. 46.1

$L = 1.57 \times R$

L = length required to form bend

Fig. 46.2 Compression bending employing a forming roller for wiping the pipe or tube around the bending form
(*Courtesy of O'Neil-Irwin Mfg. Co.*)

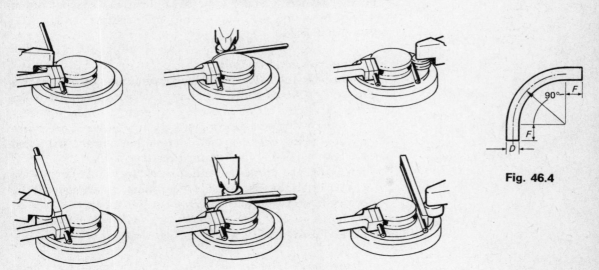

Fig. 46.4

Fig. 46.3 Compression bending using a follow block for wiping the pipe or tube around the bending form
(*Courtesy of O'Neil-Irwin Mfg. Co.*)

Compression Bending

Figs. 46.2 and 46.3 show the two techniques used and Fig. 46.4 shows the allowance.

Length of straight pipe required

$$= F + 1.57R + F$$
$$= 2F + 1.57R$$
or $\quad = 2D + 1.57R \quad \text{if} \quad D \leqq 250 \text{ mm}$

47 Applications Involving Rolling

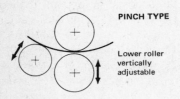

PYRAMID TYPE

Top roller vertically adjustable

Fig. 47.1

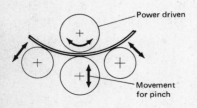

PINCH TYPE

Lower roller vertically adjustable

Fig. 47.2

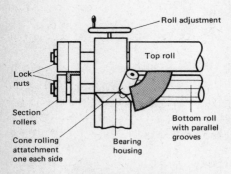

Power driven

Movement for pinch

Fig. 47.3

Roll adjustment

Top roll

Lock nuts

Section rollers

Cone rolling attatchment one each side

Bearing housing

Bottom roll with parallel grooves

Fig. 47.4 Ring rolling and cone rolling attachments

There are several types of rolls for curving thin sheet, thick plate and angle sections, but most are based on the Pyramid or Pinch type of rolls (Figs. 47.1 and 47.2).

The following is a selection of common types of rolls with typical applications.

1 Manual-Geared Bending Rolls

These may be pinch or pyramid and, for sheet metal work, grooves are cut circumferentially around the roll or at one end for bending rod and rolling wired edges. After securing, the top roller is either swung out or the top bearing housing lowered to allow rolled containers to be withdrawn. The top roller (depending on the make) may be positioned at an angle to the lower rollers to form cones. In this type of roll, capacity ranges from gauge plate to approximately 5 mm thick sheet. Parallel grooves are cut lengthways on the heavier type bottom rolls to help with the lining-up of plate edges and to facilitate scale removal.

2 Powered Bending Rolls

Again, these rolls may be of the pyramid or pinch type or a variation of these. They often have a power-driven top roller for vertical adjustment, and powered bottom rollers for rotation. For curving very heavy plate, there are often four rollers with the upper roller powered for rotation and the front and back ones supplying the curving pressure (Fig. 47.3).

Cones may be formed by adjusting the front roller or, on the pyramid type, by use of a cone rolling attachment or by sloping the top roll. For rolling sheet metal cones, a type of roll consisting of a conical-shaped top and bottom roll, open ended, is used.

Some machines have an additional attachment for rolling angle and bar sections. This attachment is an extension of the rolls, but positioned outside the bearing housing (Fig. 47.4).

Applications include structural work, boiler plates, ships frames, circular and curved tanks and containers and gas holders for heavy work; and circular sheet metal containers such as food dishes, cans and domestic ware.

An important aspect of rolling plate and sheet is the pre-forming of the ends, to obtain a true cylinder without flats where the opposite edges touch. Pinch rolls are designed to eliminate these flats at the start of the rolling operation, but other methods have to be employed when pyramid rolls are

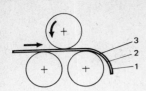

Fig. 47.5

Fig. 47.6

Fig. 47.7

Fig. 47.8

used. Briefly, some of the common methods used are as follows:

1 Using a template to check the curve, hammer the ends over the bottom roller in increments (Fig. 47.5). Thin gauge plate to approximately 6 mm.

2 Using a convex bar between the bottom roller and the edge of the plate (Fig. 47.6). Thin gauge plate to approximately 10 mm.

3 Vee block and top tool (Fig. 47.7). Thin and thick plate.

4 Pressing between blocks (hot or cold) (Fig. 47.8). Very thick plate.

When the cylinder is curved until the ends touch, it is good practice to over-roll slightly (not on very thick hot rolled plate) to allow for the elasticity in the material which "springs back" when the roll pressure is released. It is important that during rolling the weight is applied in small increments. This has the effect of "breaking in" the sheet metal and avoiding kinks in thick sheet.

There are several methods applied on powered rolls for releasing the complete cylinder: locating pin and swinging arm, counter-balanced bearing.

Safety

1 Ensure that guards are in place and that you know where the emergency stop button is for powered rolls.

2 Keep fingers clear of rollers when in motion and do not wear gloves.

3 Ensure that no-one is liable to be struck by the moving plate or that the plate falls out of the rollers.

4 Do not allow edge of plate to run through hands because of sharp edges.

Fabrication Investigation 2
Stiffness of Light Gauge Sheet

PREVIOUS KNOWLEDGE Simple calculations for joints and wiring allowances.

AIM To compare the stiffness of various forms of light gauge sheet material.

THEORY Before being formed, thin sheet metal buckles and bends when handled as well as having dangerous sharp corners and edges. By careful forming, the edges and corners can be made safe to handle and pleasant to look at. Most important of all, the strength or ability to resist being deformed is considerably increased.

APPARATUS Section holder (Fig. 47.9). Six 250 gm and six 100 gm weights and hanger. Scribing block. Millimetre rule. Various forms of fabrications as in the Table (Fig. 47.10). 230 mm × 125 mm drilled to suit section holder.

Fig. 47.9

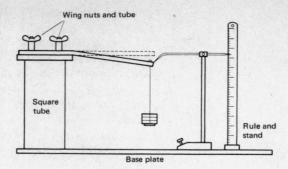

Wing nuts and tube

Square tube

Rule and stand

Base plate

PROCEDURE

After forming sections, fit the first strip with stiffened edge upwards and hang the weight hanger on the end. The bent end of the scriber must always be brought down to touch the edge and the corresponding reading taken on the rule. Hang the weights in increments of 100 gm or 250 gm and note the reading corresponding on the rule and fill in the columns in the Table (Fig. 47.10).

Plot graphs of mass against deflection.

Fig. 47.10

D = deflection in mm
W = mass in gram

Type of stiffener		W	D	W	D	W	D	W	D	W	D	W	D
Folded (single)	A												
Folded (double)	B												
False wired edge	C												
Wired edge	D												
Swaged	E												
Folded	F												
Corrugated	G												

DISCUSSION AND CONCLUSIONS

1 List the fabrications in order of their stiffness.

2 Which type of edge form would you choose for a circular mixing drum revolving at high speed and give reasons for your answer?

3 Which mechanical property allows the fabrication to return to its original shape after the weight is removed?

4 What is the main reason for corrugating roof covering sheets?

5 Take a piece of paper and fold into a corrugated sheet and support on it as many new pence as possible. Will it support one new pence without corrugating?

Fabrication Investigation 3
Comparison of Strength of Forms of Stiffening

THEORY Large areas of flat sheet, if not stiffened or supported, tend to deflect, sag or buckle. Stiffening and strengthening are achieved by supporting with simple angle or box sections, and in certain cases, in multiples of these, called compound sections. They may be joined by thermal processes such as soldering, brazing or fusion welding, or by electric resistance welding, or by riveting and bolting or self-secured joints.

Stiffness may be measured by the amount a section bends or deflects under an imposed load. The cross-sectional area and depth of a section play an important part in the resistance of a section to bending.

EQUIPMENT Number of 220 mm × 50 mm × 1 mm tinplate strips folded as in diagrams. Deflection apparatus (Fig. 47.9) and weight and hanger.

PROCEDURE
Soft solder the sections on a 220 mm × 50 mm × 1 mm strip of flat tinplate and clamp in deflection apparatus. Load with weight at one end and measure deflection, then enter results in the table (Fig. 47.11).

Fig. 47.11

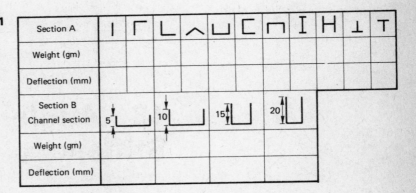

QUESTIONS
1) As the depth of the section increases, the stiffness
a) is reduced *b*) is increased *c*) remains the same.
2) Does an increase in the cross-sectional area increase the resistance to bending?
a) Yes *b*) No *c*) Remains the same.
3) By doubling the thickness of a sheet, the resistance to bending is
a) the same *b*) slightly better *c*) twice as much
d) less.

Fabrication Demonstration 1
Increasing the Stiffness and Strength of Sections

AIM To show how simple sections are built up into compound sections to increase stiffness and strength.

THEORY You will have noted from what you have just done that the depth of a section and its cross-sectional area are very important with regard to resistance to bending. If these simple sections are now combined to give compound sections, their strength and stiffness increase out of all proportion to the thickness of the section.

PROCEDURE

Using tinplate, manipulate the sections in Section A (Fig. 47.11) to make compound sections and add any others you are familiar with. One example is given below. State an application for each and show whether you would Weld (W), Bolt (B), Rivet (R) or any other method.

Section Joining Application	
Example	Box Girders
	Bolt
	Rivet
	Weld

48 Angle Frames as Stiffeners

The stiffness and rigidity of square-cornered frames may be considerably increased simply by welding, bolting or riveting diagonal members across the corners. The diagonal members also help to ensure squareness. Several of these identical units may be joined to provide very strong rigid frames. It is important from a safety point of view that diagonal members are securely fastened before lifting and erecting to prevent collapse. The two common types of welded connection are the mitre corner and the notched corner, the notched corner being more favourable as location is provided and distortion due to welding is minimised.

Sketches of these two types of corner joint are shown in Fig. 48.1.

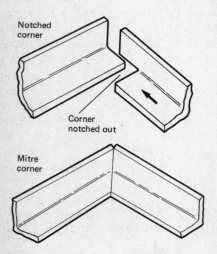

Notched corner

Corner notched out

Mitre corner

Fig. 48.1

49 Appearance, Strength and Safety in Fabricated Structural Members

When structural steelwork is designed, fabricated and erected, certain important points must be considered. The three most important points are as follows (not necessarily in order of importance).

Appearance

A fabrication should always be pleasant to look at whether it be a ship, an aeroplane, a bridge or a barn. Cambers are often introduced into structural work because they are pleasant to look at as well as providing stiffness and clearance. Sharp or abrupt angles, especially on welded beam connections, are nowadays smoothed into radii where possible, which improves both appearance and fatigue life. With the introduction of welding into highly stressed fabrications, beautiful modern designs are possible, providing smooth contours, which can also attain the highest joint strength.

Appearance is largely in the hands of the designer, but the fabricator who takes a pride in his work can certainly, by good workmanship, improve the general appearance—for example, not leaving snap and hammer marks or stray arcing on plates and sections, deslagging welds, and removing buckles. Friction grip bolts should have a uniform appearance, i.e. all heads or nuts on the same side with uniform thread. Fit-up of plates is extremely important and rivets should be uniform and not burnt. Long length flanges and stiffeners, if buckled or twisted, look most unsightly and should be avoided. Excessive drifting and straining of members should always be avoided. Surface weld defects and excessive spatter not only indicate poor workmanship and appearance, but very often that the joint strength is not of the highest.

Stanchions and trusses should be aligned to prevent sagging during assembly on a previously levelled bench area or block, and marked clearly for site erection, where required. Remember, a general tolerance for fabricated steelwork is usually ±1 mm for single dimensions, to ±2 mm for multiple dimensions.

Strength

It is said that a chain is as strong as its weakest link, and this may be said with regard to fabricated structures. The tensile strength of low carbon structural steel is 450 N/mm² and 700 N/mm² for low alloy structural steel, but it is the joint strength of the connecting members which is important. If the joint is incorrectly or badly made, either during riveting, bolting or welding, then failure may occur during service.

Butt joints and stanchion splices which transmit compressive stress should be assembled with care and accuracy to prevent unequal loading. Base gussets, angles or channels should be fixed with such accuracy that they are not reduced in thickness after machining by more than 1.6 mm. Care should be taken to ensure that the clearances specified are worked to. The erection clearance for cleated ends of members connecting steel to steel should not be greater than 1.6 mm at each end. The erection clearance at the ends of beams without web cleats should not be more than 3.8 mm at each end. If, for practical reasons, the clearance has to be increased, the seating should be suitably designed. The correct standard back marks and edge distances should be worked to unless otherwise stated.

Holes through more than one thickness, such as compound stanchions and girder flanges, should be clamped or bolted together and drilled. Punching is permitted before assembly if the holes are punched 3.2 mm less and reamed to final size after assembly. The thickness of material punched should not be more than 16 mm and all sharp burrs removed.

Care should be taken when lifting and slinging, especially with roof trusses, otherwise straining will take place and make erection extremely difficult. The sequence of erection should be correct and bracing should be safe and adequate.

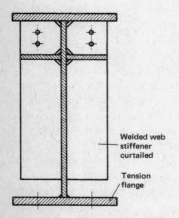

Fig. 49.1 Welded fabricated beam

Labels: Welded web stiffener curtailed; Tension flange

Web Stiffeners

As the name implies, these are (1) for making the webs of built-up members more rigid to withstand buckling and twisting due to imposed loads and (2) to transmit the load and prevent the toes of flanges buckling upwards (Fig. 49.2). The stiffeners may be welded (Fig. 49.1) or bolted or riveted (Fig. 49.2). The welding of stiffeners to a tension flange of a fabrication subject to fatigue loading is not recommended due to the lowering of fatigue life. An overhead crane and its supporting girders are typical examples of fatigue loading.

Fabrication Investigation 4
Web Stiffeners

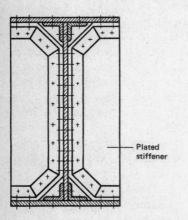

Fig. 49.2 Riveted or H.S.F.G. bolted beam.

Label: Plated stiffener

AIM To show how web stiffeners impart rigidity and strength.

EQUIPMENT   Three channel sections as shown in Fig. 49.3:
(i) no stiffeners, (ii) one stiffener central, (iii) three stiffeners
equal. Loading bar and 500 gm weights, scribing block.

PROCEDURE
Load the channel section progressively and measure the
deflection for each of the three separate positions, ensuring the
channel base is held firm in a vice. Enter the results in a Table
laid out as shown.

QUESTIONS
1)   What happens to the toes of the channel in position?
2)   What is the cause and how can this be prevented?
3)   Which channel section would you recommend for the
maximum stiffness?

Fig. 49.3

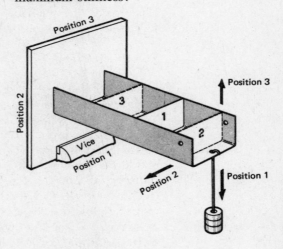

		No. 1		No. 2		No. 3	
		Weight (gm)	Deflection (mm)	Weight (gm)	Deflection (mm)	Weight (gm)	Deflection (mm)
Position 1	1						
	2						
	3						
	4						
Position 2	1						
	2						
	3						
	4						
Position 3	1						
	2						
	3						
	4						

50 Pressing

Pressing of material may be carried out in several ways: by flanging, brake or gap press (dealt with elsewhere) or by a drawing press which may be hydraulic or mechanical powered.

The press basically consists of a top punch and a bottom die and slide set within a frame. It is carefully guarded during the movement of the stroke. Often, an ejector arrangement is fitted to eject the component after pressing.

The press is used for long production runs and mass production of light metal parts such as car body components, or long-stroke heavy duty presses can be used for hot pressing of dished ends of boilers and other heavy items. Hot pressing of joggles in angles one at either end is a further example.

The Press Brake

Plain straight flanges in sheet and plate of varying degrees of angle are produced on the press brake.

A top beam, powered either mechanically or hydraulically, is brought down on to the plate which rests on a vee-shaped bottom die. The ratio of the width of this vee to the thickness of the plate being bent is important with regard to the load required to bend the plate, shown in Fig. 50.1a.

The die ratio of 8 is a practical standard to give the maximum load for the size and cost of a press. If the die ratio is less than 8, the load will be increased, with danger of the work fracturing along the bend line; if it is more, the load will be reduced. A larger die opening is often used on high tensile materials or low carbon steel above 12.8 mm to prevent cracking. Too large a die opening may cause an excess amount of metal to be drawn into the die and cause a bulge on the outside radius.

The radius on the upper die should not be less than the metal thickness being bent, and is usually 2 to 3 times T.

The dies are bevelled to 85° to enable plates to be bent readily to 90° after springback of the plate when the load is released. A dial is often fitted on the heavier plate pressing machinery to indicate the pre-set forming depth. This dial saves frequent checking with the angle bending template. Top dies of different shapes including rounds as well as vees, are used to produce components.

A cranked tool (Fig. 50.1b) is often used for forming box sections.

Another feature of the press brake is that punching, blanking and piercing of sheet may be carried out using special top tools and dies.

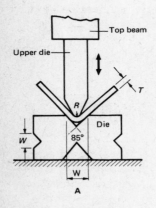

A

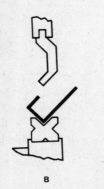

B

Fig. 50.1

Safety

Care of sharp edges; beware of crushed fingers. Ensure that lifting tackle is secure and that work does not fall after top beam returns. Watch out for movement during bending and avoid thumb or fingers being crushed between plate and machine body.

Note: Most modern press brakes are now fitted with a photo-electric guard.

Fabrication Investigation 5
V-factor

AIM   To prove the width of V-factor using various thicknesses of material and also different widths of Vee.

EQUIPMENT   15 ton hydraulic press (or suitable alternative).

Three bottom dies of different vee widths to suit metal thickness, and top dies.

Low carbon steel plate, 1.6 mm, 3.2 mm, 6.4 mm, three off each, all the same length and marked with a centre line.

PROCEDURE

For each die, press one of each plate thickness on the centre line and note the results.

51   Spinning and Drawing

Spinning

This rotary forming process produces parts which are round in cross-section with sides contoured into cylinders, cones or domes. Relatively thin metal is used to produce the shape of the finished blank.

The process consists of the cold plastic deformation of material to the shape of a rotating mandrel by means of a hand held tool or a power fed roll (Fig. 51.1). Lubricants are used for different metals to assist with the forming and care must be taken on radii not to corrugate or thin the metal and cause tearing. By using special tools the edges may be beaded to give a rigid safe edge.

The important points to note about spinning are:

1   The process causes work hardening of the component.

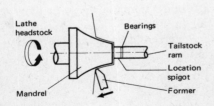

Fig. 51.1

2   Only metals which are fairly ductile may be spun successfully.

3   The thickness of the metal is not altered intentionally.

4   The surface area of the finished article is the same as that of the initial blank.

Safety

1   Ensure that the blank disc is held securely.

2   Check that the tailstock ram is in the correct position and exerting the correct pressure.

3   Do not finger the revolving disc and stand to the side.

4   Keep the tool below dead centre when manual spinning.

Drawing

This is a process where sheet metal parts are pressed into various shapes using a top and bottom die and either hydraulic, pneumatic or mechanical power. A blank piece of metal, brass, steel, copper, aluminium or stainless steel, is placed on the bottom die. The top die (which fits into the bottom die) is brought down to press out the desired shape and in so doing, the metal from around the edge is *drawn* in between the dies. If the depth of the die is relatively deep, then the process is termed "deep drawing". Metals must be both strong and ductile for deep drawing. Examples are shown in Fig. 51.2.

Some modern drawing presses have a hydraulic oil pressure chamber with the bottom closed by a strong, flexible diaphragm as an upper unit. When the upper unit is brought into contact with the bottom die and blank, an oil pressure is built up so that the diaphragm takes the shape of the bottom die and helps to eliminate tears and wrinkles. The result is a very smooth finish. The blank piece of metal may have a thin film of plastic adhering to prevent marking and assist the drawing operation.

The important points to note about drawing are:

1   For the drawing operation to be successful, the metal must pass its yield point to enable plastic flow to take place.

2   When deep drawing certain articles, more than one forming operation is required to avoid excessive wrinkling and splitting.

3   Some drawing operations require that the dies are lubricated to allow free "run in" of the blank.

4   The type and quality of the material has a great bearing on the quality of the finished product.

Safety

1   Keep fingers clear during drawing operation.

2   Do not remove guards.

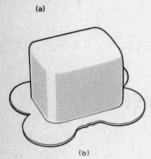

(a)

(b)

Fig. 51.2

52 Riveted and Bolted Joints

Fig. 52.1 shows some common types of joint which may be bolted or riveted.

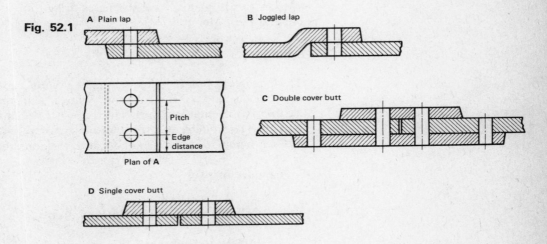

Fig. 52.1

A Plain lap

B Joggled lap

Pitch
Edge distance

Plan of A

C Double cover butt

D Single cover butt

Riveted Joints

Riveted joints should have all parts firmly drawn and held together before and during riveting with a service bolt in every third or fourth hole. When hot riveting is carried out, the rivets should be heated uniformly along their length without burning or excessive scaling, so as to provide a head of standard dimensions. After being driven, they should completely fill the hole and, if countersunk, the countersinking should be fully filled and any proudness of the countersunk head dressed flush if required. Machine riveting of the steady pressure type should be used wherever possible and any loose, burned or defective rivets should be removed and replaced.

Where both riveting and welding are used in making a connection, the welding should be completed first.

Low carbon steel rivets, especially when hot, are extremely malleable and plastic, hence their ability to be closed to form joints in steel structures and sheet.

A correctly made riveted joint should

a) have the shank swelled to fill the hole;

b) have sufficient material projecting through the hole to just form a satisfactory head;

c) have the heads in alignment;

d) not have tool marks around the head;

181

e) not be buckled due to too much pressure from the riveting operation;

f) not have "burnt" rivets;

g) have all holes drilled or punched and reamed and all drilling burrs removed with a counter-sinking tool;

h) if possible, have all plates brought together, then drilled through (solid drilling);

i) not have holes formed by gas cutting process;

j) on curved plates, have the holes drilled for riveting after curving the plate.

Clearance

The diameter of rivet holes is usually 1.6 mm larger than the nominal diameter of the cold rivet (above 12.6 mm dia.). The clearance for smaller rivets which are to be closed cold is normally just sufficient to enable the shank to enter the hole without hammering. One method for calculating the clearance is

Diameter of rivet $\times 0.05$

Some Common Defects in Riveted Joints

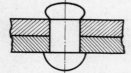

A) Insufficient shank projecting through hole to form full head (Fig. 52.2).

Causes

Rivet length too short.

Hole oversize.

Snap cup too large.

Rivet not hot enough when hot riveted.

Fig. 52.2

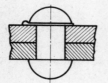

B) Excess shank projecting causing a "Jockey Cap" (Fig. 52.3).

Causes

Rivet length too long.

Snap out of line.

Rivet bent over out of line.

Fig. 52.3

C) Heads not in line (Fig. 52.4)

Causes

Accumulated errors causing holes out of alignment.

Snaps not in line.

Rivet inserted at an angle.

Fig. 52.4

D) Shank not swelled in hole (Fig. 52.5).

Causes

Plates not bolted tight.

Striking side of rivet before swelling shank by striking on centre.

Fig. 52.5

Fig. 52.6

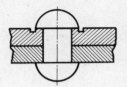
Cup head

E) Springing of plates between holes (Fig. 52.6).
Causes
Not enough tack bolts.
Excessive drifting.
Not holding riveting pressure on long enough when hot riveting.

Fig. 52.7

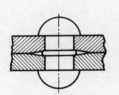

F) "Snap" or hammer marks on plate (Fig. 52.7).
Causes
Continuing riveting too long.
Holding snap at a wide angle during riveting.
Rivet too short.

Fig. 52.8

G) Spread of shank in between plates (Fig. 52.8).
Causes
Lack of tack bolts.
Bolts not tight.
Plates not closed down before spreading rivet shank.

53 Types of Rivet and Rivet Head

Fig. 53.1

Snap head

Snap Head (Fig. 53.1)
Easily shaped by machine or hand snaps. Sometimes called a cup head after forming.
Applications Structural work such as road, rail and river bridges, steel framed buildings, ducting and boilers.

Fig. 53.2

Pan head

Pan Head (Fig. 53.2)
May be "held up" during riveting by a flat dolly or jamback and used where maximum strength is required.
Applications Locomotive and diesel bogie frames. Foundations and grillages.

Fig. 53.3

Rounded countersunk

Rounded Countersunk (Fig. 53.3)
(Countersunk Cup)
Used where clearance is necessary but not required to be flush.
Applications Boiler work—such as foundation rings and fire hole plates.

Fig. 53.4

Flat countersunk

Flat Countersunk (Fig. 53.4)
Used where flat surfaces are required and joining thick plates to thin.
Applications Chutes, ducts, hoppers, ships decks and rails.

Mainly for Sheet Metal

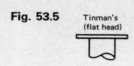

Fig. 53.5   Tinman's (flat head)

Tinman's (Flat Head) (Fig. 53.5)
Used for sheet metal joints where least obstruction is desirable. Flat dollies may be used to "hold-up" the rivet.
Applications   Hoods, ventilating ducts, hoppers and panels.

Pop Rivet (Fig. 53.6)
There are many other kinds of rivet used for joining sheet metal but the pop rivet is one of the most popular. It is fitted into the drilled hole and formed either by using lazy tongs or a plier type tool.
Applications   Assembly of light fabrications, vehicle panels, ductwork and containers.

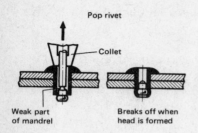

Pop rivet

Collet

Weak part of mandrel

Breaks off when head is formed

Fig. 53.6

Examples of Rivets used in Sheet Metal Fabrication
Fig. 53.7 shows semi-tubular and full tubular rivets before and after setting.

Fig. 53.8 shows a blind rivet, which are ideal for limited access installations. Minimum back-up clearance is needed.

Fig. 53.9 shows a rivet set by a single hammer blow.

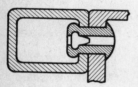

(a)   (b)

Fig. 53.7 (a) Semi-tubular, (b) full tubular

Forms of Driven Head

Knobbled or Conical (Fig. 53.10)
Applications   Used on hoppers and chutes to reduce obstruction.

Used on shipbuilding, structural and locomotive and diesel work for repairs.

Hand hammers and flat snaps are used where it is not possible for machine riveting to be carried out.

Fig. 53.8   Blind rivet

Cup Head
Applications   As for snap head. The cup may be varied in depth depending on the type of snap used.

Fabrication Investigation 6
Rivet Shanks and Cap Volume

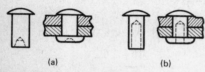

Fig. 53.9

AIM   To investigate the length of rivet shank projecting through the plates and compare with the cap volume of the snap.

THEORY   The amount of rivet shank projecting should be equal to the volume of the cup which is to form the head, plus the amount required to be pushed back to fill the clearance around the hole. The clearance and length of holes varies so it is always advisable to make a trial rivet before cutting rivets to size. The amount required varies with thickness of plates and cup size, and of course, for countersunk and knobbled heads (from 0.8 to $1.5 \times$ diameter for countersunk holes, and $2.0 \times$ diameter for snap heads).

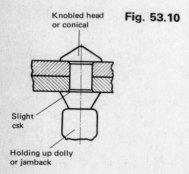

Knobled head or conical   **Fig. 53.10**

Slight csk

Holding up dolly or jamback

	Dia. of Rivet and Snap Size	Length of Plasticine Cylinder	Approx. Length Projecting	Final Length Projecting	Remark
	mm	mm	Trial length		
Example 1	4.8	7.2	1.5 × dia. of rivet	2 × dia. of rivet	Final length, perfect head.
2					
3					
4					
5					
6					

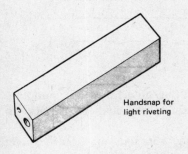

Handsnap for light riveting

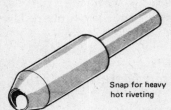

Snap for heavy hot riveting

Fig. 53.11

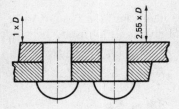

Fig. 53.12

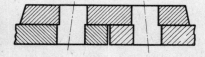

Fig. 53.13

EQUIPMENT Various sizes of rivet snaps with cup end. Snap head rivets to suit. Plasticine.

PROCEDURE
Fill each cup with plasticine and level off.

Remove plasticine and roll into a cylinder the same diameter as the rivet which suits the snap.

Complete the Table and the Example.

Example
Lap joint between two 6.4 mm plates, holes drilled 8 mm dia., rivets 7.2 mm dia. (d). Amount finally projecting, 14.4 mm ($2 \times d$). Method of riveting, vertical hydraulic press, hot rivets. Result, a perfect cup.

Fabrication Demonstration 2
Forming the Driven Head: Amount of Material

AIM Incorrect amount of metal for forming the driven head.
EQUIPMENT AND MATERIAL Plates and rivets to suit snaps available (Fig. 53.11), clearance drill, riveting equipment, hand or machine.

PROCEDURE
Complete the riveting, then draw the result of the formed heads on Fig. 53.12. Compare the result with rivet defects and state the type of joint.

Fabrication Demonstration 3
Forming the Driven Head: Misalignment

AIM Misalignment of the driven head. Holes out of alignment (Fig. 53.13).

PROCEDURE
As for Demonstration 2.

Fabrication Demonstration 4
Accumulated Errors

THEORY These are usually caused by incorrect setting of dividers, causing each small error to be added to the next progressively, or by marking off each single dimension in a line of holes. The correct procedure is to mark off the holes as an additive total, shown in Fig. 53.14, known as "string dimensions".

MATERIAL Two sheets, 260 mm long, dividers, scriber, centre pop or prick punch, ruler, drill.

Fig. 53.14

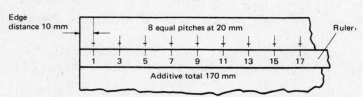

EXAMPLES

1   Set pair of dividers to 20 mm plus an error of 0.05 mm and mark off twelve equal pitches and note the final error (mark on sheet).

2   Using the "string dimension" method, mark off twelve equal 20 mm pitches and check each for accuracy.

3   Drill holes and bolt plates together and sketch the results as in Fig. 53.15.

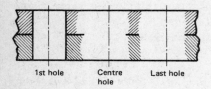

Fig. 53.15

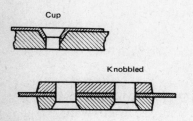

Fig. 53.16

Fabrication Demonstration 5
Countersunk Riveting of Thin Material to Thick

MATERIALS As for Demonstration 3 plus flat back up dolly and 60° countersinking bit.

PROCEDURE

After riveting, sketch in the driven heads and the type of joint as in Fig. 53.16. Sketch the type of riveting tool used.

54   Types of Bolt

Black Bolt

Usually hot forged and roll or machine threaded. Used mainly as a *service bolt,* for temporarily bolting fabrications together prior to riveting and for tack bolting. They may also be used as permanent bolts for lightly stressed fabrications. They are the cheapest form of bolt available.

Clearance The diameter of the bolt hole is usually 1.5 mm larger than the nominal diameter of bolt shank (unthreaded portion).

Close Tolerance Bolt

These are bolts which have been machined under the head and on the shank to give a more accurate finish. They are used where compound members of several thicknesses require accurate alignment and where it is essential that no movement of the work must take place prior to riveting.

The clearance is as for Turned Barrel Bolts.

Holes are finally reamed to correct size.

Turned Barrel Bolt

These bolts are fully machined with the diameter of the screwed portion of the bolt being 1.5 mm smaller than the diameter of the barrel, to prevent damage to the threads when fitting. The barrel length should bear fully on all parts connected with no thread in the hole (Fig. 54.1). These bolts are used when the highest accuracy is required, and sub-assembly of fabrications prior to checking for alignment, and fabrications subject to heavy loads.

Fig. 54.1 Correctly bolted joint

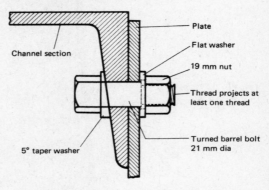

Plate
Flat washer
19 mm nut
Thread projects at least one thread
Turned barrel bolt 21 mm dia
Channel section
5° taper washer

Clearance Holes are drilled and reamed to a diameter equal to the nominal diameter of the barrel subject to a tolerance of +0.13 mm and −0 mm. (The bolt diameter is 0.13 mm smaller than the hole.)

High Strength Friction Grip Bolts (H.S.F.G.) (BS 4395 and 4604)

There are several types of H.S.F.G. bolt (Fig. 54.2), but all must be tightened to the specified minimum tension.

Fig. 54.2 High strength friction grip bolts

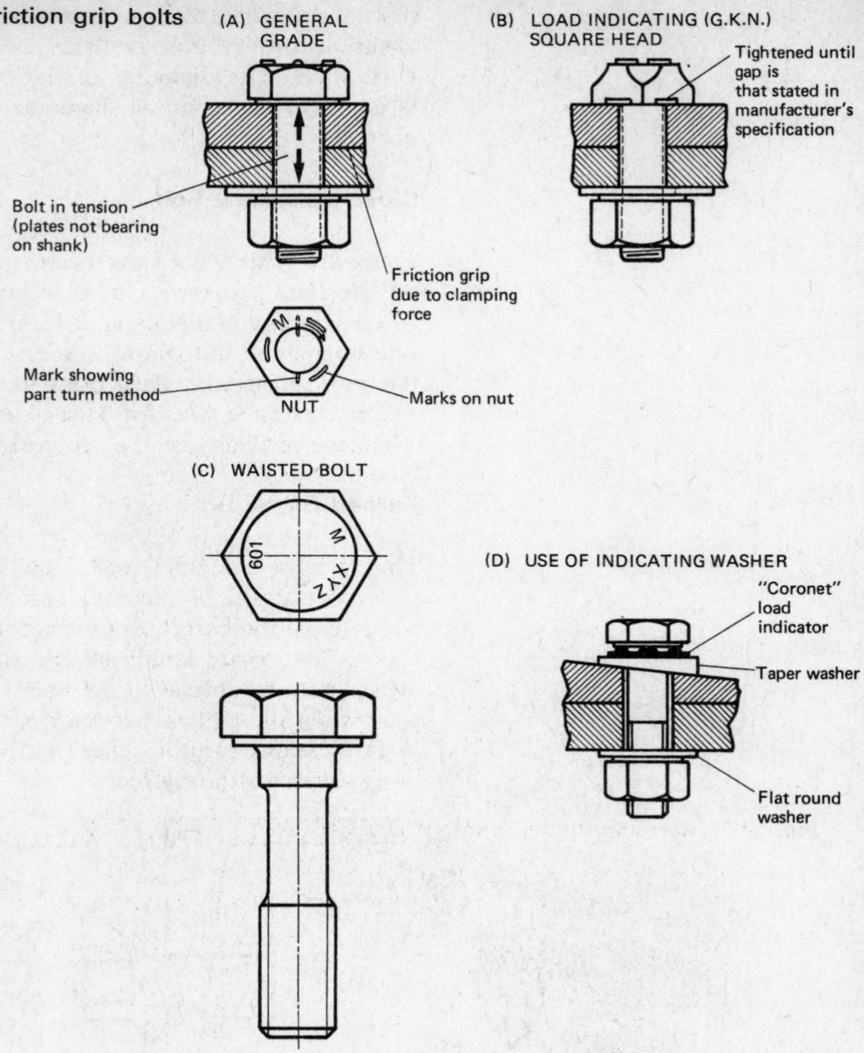

(A) GENERAL GRADE

Bolt in tension (plates not bearing on shank)

Friction grip due to clamping force

Mark showing part turn method

NUT

Marks on nut

(B) LOAD INDICATING (G.K.N.) SQUARE HEAD

Tightened until gap is that stated in manufacturer's specification

(C) WAISTED BOLT

(D) USE OF INDICATING WASHER

"Coronet" load indicator

Taper washer

Flat round washer

Fig. 54.3 Bolt head
ISOM = ISO metric identification
 XYZ = manufacturer's (trade) marking
 601 = strength grade
      (see Fig. 54.2C)

The term H.S.F.G. relates to bolts of high tensile steel, used in conjunction with high tensile steel nuts and hardened steel washers. They are tightened to a predetermined shank tension in order that the clamping force thus provided will transfer loads in the connecting members by friction between the parts and not by shear in, or bearing on, the bolts or plates of connecting members.

These bolts are made by a special cold working process which includes two operations: heading and thread rolling. Fig. 54.3 shows the head for identification. Close tolerances ensure accuracy, and heat treatment is carried out afterwards.

The surfaces in contact must be free from paint, oil, dirt, loose rust and scale. Each bolt is assembled with one washer in cases where plane parallel surfaces are involved. The washer is placed under the bolt head or nut, whichever is to be rotated during the tightening operation. A tapered washer must be used also if the angle is above 3°.

Driving of bolts is not permitted.

If, after final tightening, a nut or bolt is slackened off, it must *not* be used again.

Higher grade (waisted shank) bolts BS 4604 may be used in joints subject to tension as well as shear, using the part turn method of tightening. Tightening by the torque control method is not permitted.

Applications and Advantages

Used on road bridges, structural repairs and extensions, heavy installations subject to vibration, power station work and colliery winding gear.

H.S.F.G. bolts have virtually replaced hot driven rivets for fastening steel structures on site. They are also used to replace defective rivets on repair work.

Where structural members are shop assembled, welding is still more economical on the whole, but certain connections are better bolted and the difficulties encountered with site welding are eliminated when H.S.F.G. bolts are used.

System-built factories and offices are particularly suitable for H.S.F.G. bolts.

Compared with riveting, fewer H.S.F.G. bolts are needed than rivets for a given joint, and fewer men in the team needed to put them in, two against three or four. Less noise, simpler assembly (approximately three times faster than riveting), simpler inspection, and no fire risk. A smaller safety factor may be allowed than for riveted structures. On the whole, the cost is less using H.S.F.G. bolts.

Note: It is important that the torque on the nuts is correct for the bolt, so a pre-calibrated impact wrench is used, or the part turn method, or a feeler gauge if load indicating bolts or washers are being used. Bolts must be tightened in a definite sequence.

Clearance The diameter of the bolt hole is usually 1.6 mm larger than the nominal diameter bolt shank.

Washers

General

In all cases where the full bearing area of the bolt is to be developed, the bolt should be provided with a steel washer under the nut. This washer must be of sufficient thickness to avoid any threaded portion of the bolt being within the thickness of the parts bolted together and to prevent the nut, when screwed up, bearing on the shank of the bolt.

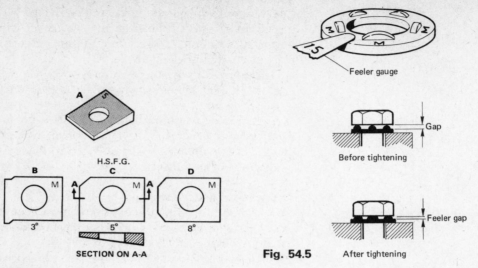

H.S.F.G.

SECTION ON A-A

Fig. 54.4

Feeler gauge

Gap

Before tightening

Feeler gap

Fig. 54.5   After tightening

Taper (Fig. 54.4)
These washers should be placed under the nut or head of any bolt bearing on any bevelled surface. The angle of taper on the washer should match the angle of taper on the section. Washers are invariably marked with the angle of taper, i.e. 3°, 5° or 8°, or with the type of corner on H.S.F.G. taper washers (types B, C, D in Fig. 54.4).

Load Indicating Type "Coronet" (Fig. 54.5)
The "Coronet" Load Indicator is a specially hardened washer with protrusions on one face. The protrusions bear against the underside of the bolt head leaving a gap. As the bolt is tightened the protrusions are flattened and the gap reduced. At a specified average gap, measured by feeler gauge, the induced shank tension will not be less than the minimum required by Standards.

Fig. 54.6

Spring Types (Fig. 54.6)
Where vibration is likely to be present to any degree in a fabrication, the nuts should be used with some form of spring washer to prevent the nut unscrewing during service.

Flat Type (Fig. 54.6, 54.7)
Ordinary flat, low carbon steel washers are just punched into shape without further work but washers for turned barrel bolts are machined and should have a hole diameter not less than 1.6 mm larger than the barrel thickness and a thickness of not less than 3.2 mm so that the nut will not bear on the shoulder of the bolt. Flat washers for H.S.F.G. bolts are made from low alloy steel and are extremely hard and may be distinguished by the three tabs on the outer edge (Fig. 54.7).

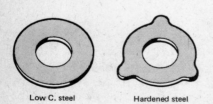

Low C. steel   Hardened steel

Fig. 54.7

190

55 The Use of Templates for Fabricated Assemblies

Depending upon the type and application of template required, various materials are used. The following is a selection of materials with a typical application and reasons for using that material.

1   *Material:*   Red/green template paper.
     Applications: Developed pipes, chutes, ducts and bends, cones, angles and plates.
     Reasons: Easily marked, use of both sides, flexible for wrapping around pipes, etc., cuts easily, comparatively cheap and can be used again.

2   *Material:*   Strawboard (brown lined, double sided).
     Applications: Structural gusset plates, shoe plates, holes and back marks in angles, tees and joists. Profile cutter templates. The heel intersections for purlins and rafters on roof trusses.
     Reasons: Easily marked with pencil, may be placed on section and punched through, can be cut on one side to form a hinge. Stiffer than paper (cannot be obtained in small amounts). Reasonably cheap in large quantities.

3   *Material:*   Whitewood battens.
     Applications: Marking holes, mitres, notches, back marks and cross centres on beams, stanchions and girders. Profile followers (page 121). Box templates.
     Reasons: Easily marked and seen, either by pencil or paint, easily cut and machined, light in weight, cheap and may be reclaimed and planed down.

4   *Material:*   Metal (thick or thin).
     Applications: All classes of structural or developed work, either as long lasting templates or scheduled to become part of the fabrication. Magnetic profile templates (see p. 121).
     Reasons: Strong, tough, material able to withstand shop floor usage and storage. May be used for locating drills and nipple punching by hand.

When large quantities of templates are required, a template loft with a close jointed, black, diagonally boarded floor is used. For structural work, the temperature should be a constant 20°C in the loft.

Templates are used to carry information from the loft to the shop floor, very often in the form of symbols and colour codes. British Standard back marks and edge distances and cross centres are worked to unless otherwise stated.

Examples of information carried on templates are as follows:

1 Number off required, e.g. 12 off.
2 Mark for identification or assembly, e.g. Mk 2A.
3 Job Number, e.g. WK/28/232.
4 Hole diameters may be denoted by symbols, e.g. ⊙ 15 mm, ⊡ 19 mm, ▽ 21 mm, ◈ 24 mm.
5 Cutting lengths, e.g. 299 mm

6 Slope required, e.g.

7 Bending instructions, e.g.

    / Bend \ Down

8 S.O.P. = Setting Out Point.

From the drawing shown in Fig. 55.1 mark out the template for the gusset and add all the relevant information on the template.

56 Practical Developments in Sheet Metal

Right Cylindrical Tee Piece on Centre (with flange)

1 Using the parallel line method of development, draw a full size set out on template paper the developments as shown in Fig. 56.1. Write out the sequence of operations for complete fabrication. (See examples page 197.)
2 If time permits, mark out on plate and fabricate, then check for dimensional accuracy. See page 166.
3 As an alternative, right-angled tee pieces between squares or rectangular pipes of equal size may be developed using the parallel line method.

Fig. 55.1

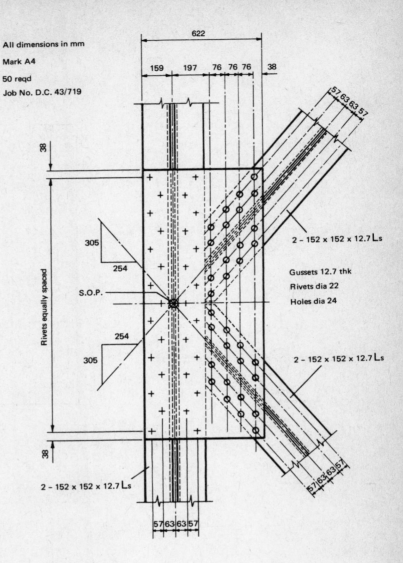

All dimensions in mm

Mark A4

50 reqd

Job No. D.C. 43/719

Tin Box

1   Draw the developments shown in Fig. 56.2 on template paper. The sequence of operations is given for manufacture in tin plate.

The Right Cone (Tundish)

1   Using the radial line method of development, draw a full size set out on tinplate the developments shown in Fig. 56.3. The sequence of operations required for complete fabrication is given.

2   Now fabricate the tundish.

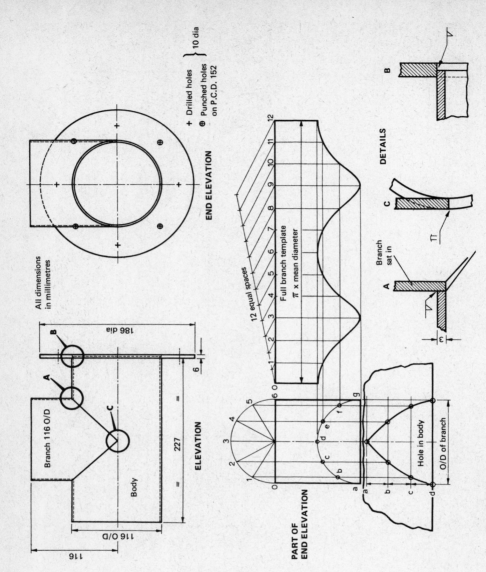

Fig. 56.1 Tee piece

END ELEVATION

+ Drilled holes
⊕ Punched holes on P.C.D. 152

} 10 dia

All dimensions in millimetres

ELEVATION

Branch 116 O/D

Body

186 dia

116 O/D

116

227

6

DETAILS

B

C

A

Branch set in

PART OF END ELEVATION

12 equal spaces

Full branch template
π × mean diameter

Hole in body

O/D of branch

194

A typical industrial application of cylindrical tee pieces (*Courtesy International Combustion Ltd.*)

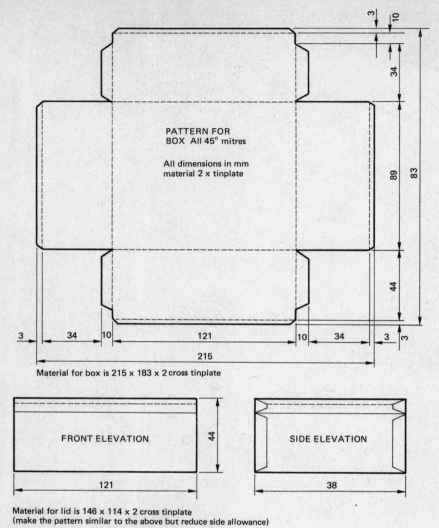

PATTERN FOR
BOX All 45° mitres

All dimensions in mm
material 2 x tinplate

3
10
34
89
83
44
3
34
10
121
10
34
3
3
215

Material for box is 215 x 183 x 2 cross tinplate

FRONT ELEVATION

44

121

SIDE ELEVATION

38

Material for lid is 146 x 114 x 2 cross tinplate
(make the pattern similar to the above but reduce side allowance)

Fig. 56.2 Tin box

Operation Sheet for Sandwich Tin Box in Tin Plate

	Operation	Tools Used	Safety
1	Check tin plate for overall length and width.	Rule.	Sharp edges.
2	Remove any unwanted material.	Treadle guillotine.	
3	Mark out development of box.	Rule, dividers, square.	
4	Check marking-out.	Rule and square.	
5	Cut out development.	Tin snips.	
6	Check corners for squareness.	Square.	
7	Fold development into box.	Folding machine.	Beware of trapping
	a) safety edge	Folding machine.	fingers and counterweight.
	b) ends including tabs	Box or pan folding machine, hide mallet.	
	c) sides.		
8	Hold tabs in position and soft solder.	Flux, solder, soldering iron.	Beware of flux and hot soldering iron, especially on face and eyes.
9	Wash off surplus flux.		
10	With tin plate supplied, mark out development for lid.	Rule, square and dividers.	
11	Cut out development.	Tin snips.	
12	Check development.	Rule and square.	
13	Fold development into lid	Bench folding machine.	
	a) safety edge	Folding machine.	
	b) ends including tabs	Box or pan folding machine, hide mallet.	
	c) sides		
14	Solder tabs.	Flux, solder, soldering iron.	Beware of flux and hot soldering iron.
15	Wash off surplus flux.		
16	Fit lid to box.		
17	Polish finished sandwich tin.	White powder cloth.	

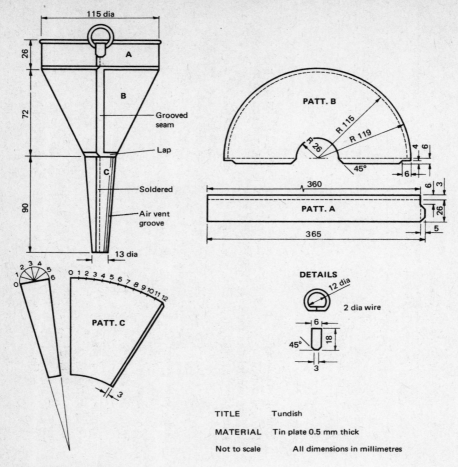

115 dia

26

A

72

B

Grooved seam

Lap

Soldered

C

90

Air vent groove

13 dia

PATT. B

R 26
R 115
R 119
45°
4
6
6

360
PATT. A
6
3
26
365
5

PATT. C

0 1 2 3 4 5 6
1 2 3 4 5

O

0 1 2 3 4 5 6 7 8 9 10 11 12

3

DETAILS

12 dia
2 dia wire

6
45°
18
3

TITLE   Tundish

MATERIAL   Tin plate 0.5 mm thick

Not to scale   All dimensions in millimetres

Fig. 56.3 Tundish

Operation Sheet for Tundish in Tin Plate

	Operation	Tools Used	Safety
1	Mark out pattern A to dimensions shown on drawing.	Rule, square, dividers, scriber.	Sharp edges.
2	Cut out shape required.	Tin snips.	
3	Cut length of 13 s.w.g. wire and square off ends.	Wire cutters and file.	
4	Straighten wire.	Hammer.	
5	Fold over allowance for wired edge, insert wire and form wired edge.	Bench folding machine, hide mallet.	
6	Form material into required ring.	Bench rolls.	
7	Solder seam.	Flux, solder, soldering iron.	Beware of flux and hot soldering iron.
8	Mark out pattern B.	Rule, square, dividers, scriber	
9	Check marking out.		Sharp edges.
10	Cut out pattern.	Tin snips and small curved snips.	
11	Break grain of material (tin plate).	Bench rolls.	
12	Prepare edges to form grooved seam.	Hatchet stake, hide mallet.	
13	Form pattern into funnel body.	Funnel stake.	
14	Insert both halves of joint and form grooved seam.	Funnel stake, grooving tool hammer.	
15	Solder seam inside of funnel.	Flux, solder, soldering iron.	Beware of flux and hot soldering iron.
16	Round up funnel.	Funnel stake.	
17	Turn top edge of funnel up.	Jenny machine.	
18	Fit top ring to funnel body.		
19	Solder in position.	Flux, solder, soldering iron.	Beware of flux and hot soldering iron.
20	Mark out Pattern C for spout	Rule, scriber, dividers.	
21	Check marking out.	Rule.	Sharp edges.
22	Cut out pattern.	Tin snips and curved snips.	
23	Form pattern into spout shape.	Tapered stake, hide mallet.	
24	Solder seam.	Flux, solder, soldering iron.	Beware of flux and hot soldering iron.
25	Round up spout.	Stake and mallet.	
26	Put in air exhaust groove.	Stake, hammer and shaping tool.	
27	Fit spout to body.		
28	Solder in spout.	Flux, solder, soldering iron.	Beware of flux and hot soldering iron.
29	Locate and fix in position ring and tab supplied.	Flux, solder, soldering iron.	Beware of flux and hot soldering iron.
30	Polish finished tundish.	White powder and cloth.	Beware of flux and hot soldering iron.

General Questions

1. Workshop safety is the responsibility of:
   (a) employers only   (b) employees only
   (c) employers and employees
   (d) the work's committee
2. The maximum lifting capacity shown on lifting equipment is usually prefixed by the letters:
   (a) M.W.L.   (b) T.W.L.
   (c) S.W.L.   (d) H.W.L.
3. Sling chains used for lifting are required by law to be inspected:
   (a) once every 5 years   (b) twice per year
   (c) once per month   (d) twice per month
4. The live wire to a three point electric plug is coloured:
   (a) black   (b) blue
   (c) green   (d) brown
5. The safest type of machine guard is:
   (a) fixed   (b) adjustable
   (c) hinged   (d) telescopic
6. On acetylene gas systems left-hand threads are used:
   (a) throughout   (b) on cylinders only
   (c) on hoses only   (d) on torches only
7. The colour code for identifying a gas cylinder containing propane is:
   (a) maroon   (b) black
   (c) blue   (d) red
8. Oil and grease can ignite violently when in contact with pressurised:
   (a) carbon dioxide   (b) propane
   (c) oxygen   (d) hydrogen
9. A person suffering from the effects of electric shock should be:
   (a) given a cold drink
   (b) given an alcoholic drink
   (c) encouraged to walk
   (d) kept warm and covered
10. An electric fire should be attacked by:
    (a) spraying with water
    (b) disconnecting the electrical connection
    (c) using a CO_2 gas extinguisher
    (d) covering with sand
11. The rest on a pedestal grinding machine is adjustable to allow for the:
    (a) size of work piece
    (b) grinding of bevels
    (c) thickness of the wheel
    (d) wear of the wheel

12. Using only a steel tape the squareness of a rectangular template may be checked by measuring and comparing the length of:

(*a*) one diagonal and one long side

(*b*) one diagonal and one short side

(*c*) two parallel sides

(*d*) the two diagonals

13. A worker is required to establish a horizontal datum. Which of the following tools would he use?

(*a*) compasses (*b*) spirit level

(*c*) trammels (*d*) sine bar

14. When marking out, what is the primary line called from which the position of other details are fixed?

(*a*) datum line (*b*) plumb line

(*c*) base line (*d*) main line

15. The distance from centre to centre of holes on the same line is called the:

(*a*) backmark (*b*) step

(*c*) edge distance (*d*) pitch

16. If low carbon steel is welded using a carburising flame, the joint will be:

(*a*) brittle (*b*) oxidised

(*c*) soft (*d*) satisfactory

17. Which of the following cutting processes is classified as non-chip forming?

(*a*) shearing (*b*) grinding

(*c*) sawing (*d*) drilling

18. Which gas combined with oxygen gives the highest flame temperature?

(*a*) natural gas (*b*) hydrogen

(*c*) propane (*d*) acetylene

19. Nibbling machines are used for:

(*a*) producing smooth surfaces

(*b*) cutting out complex shapes

(*c*) cutting long straight edges

(*d*) producing edge preparations for welding

20. Regulators are used on high pressure oxy-acetylene welding systems to:

(*a*) reduce cylinder pressure to working pressure

(*b*) increase the gas velocity

(*c*) control the temperature of the flame

(*d*) purify the gas before reaching the torch

21. The blowpipe flame suitable for oxy-acetylene welding of low carbon steel is:

(*a*) oxidising (*b*) carburising

(*c*) neutral (*d*) carbonising

22. Which type of flame is suitable for oxy-acetylene welding brass?

(*a*) carburising (*b*) oxidising

(*c*) neutral (*d*) reducing

23. The welding fault which cannot be identified by simple visual inspection is:

(*a*) lack of penetration   (*b*) unfilled crater

(*c*) undercut   (*d*) lack of inter-run fusion

24. Which of the following is a weld defect?

(*a*) undercut   (*b*) heat-affected zone

(*c*) penetration bead   (*d*) fusion zone

25. One function of the coating on a manual metal-arc welding electrode is to:

(*a*) prevent rusting

(*b*) increase welding current

(*c*) stabilise the arc

(*d*) reduce arc temperature

26. The metal arc welding process is a type of:

(*a*) pressure welding   (*b*) resistance welding

(*c*) fusion welding   (*d*) forge welding

27. The purpose of a welding rectifier is normally to:

(*a*) convert d.c. to a.c.

(*b*) correct current changes

(*c*) correct voltage changes

(*d*) convert a.c. to d.c.

28. An advantage of a d.c. welding plant over an a.c. plant is that one can:

(*a*) use high voltages   (*b*) use lightweight cable

(*c*) change polarity   (*d*) use high amperages

29. Which of the following will reduce the mains voltage suitable for *arc welding*?

(*a*) transformer   (*b*) fuse box

(*c*) rectifier   (*d*) motor generator

30. Porosity is most likely to be caused by:

(*a*) damp electrodes

(*b*) incorrect electrode manipulation

(*c*) too low a current value

(*d*) too high a current value

31. The weld defect known as undercut is caused by:

(*a*) too great a heat build up

(*b*) too low a current value

(*c*) damp electrodes

(*d*) insufficient cleaning and preparation

32. Gas entrapped within a weld is known as:

(*a*) slag inclusion   (*b*) lack of penetration

(*c*) blow back   (*d*) porosity

33. Complete the following table of identification

Gas	Colour of Cylinder
Acetylene	
Propane	
Argon	
Oxygen	

Recommended Dimensions for Back Marks and Hole Cross Centres

RECOMMENDED SPACING OF HOLES IN COLUMNS, BEAMS AND TEES TO BS 4:Part 1:1972

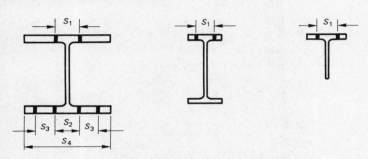

Nominal flange width	Spacing of holes				Maximum dia. of rivet or bolt	b min
	S_1	S_2	S_3	S_4		
mm	mm	mm	mm	mm	mm	mm
419 to 368	140	140	75	280	24	362
330 and 305	140	120	60	240	24	312
330 and 305	140	120	60	240	20	300
292 to 203	140				24	212
190 to 165	90				24	162
152	90				20	150
146 th 127	70				20	130
102	54				12	98
89	50					
76	40					
64	34					
41	30					

Note: That the actual flange width for a universal section may be less than the nominal size and that the difference may be significant in determining the maximum diameter.

The dimensions S_1 and S_2 have been selected for normal conditions but adjustments may be necessary for relatively large diameter fastenings or particularly heavy masses of serial size.

b min. This is the minimum width of flange to comply with Table 21 of BS 449: Part 2: 1969

RECOMMENDED BACK MARKS
FOR HOLES IN CHANNELS
TO BS 4:Part 1:1972

Nominal flange width	S_1	Maximum dia. of bolt or rivet
mm	mm	mm
102	55	20
89	50	20
76	45	20
64	35	16
51	30	10
38	22	

Symbols for Rivets and Bolts

Open Holes	
Open Holes Csk. near side	
Open Holes Csk. far side	
Shop Rivets	
Shop Rivets Csk. near side	
Shop Rivets Csk. far side	
Shop Bolts	
Shop Bolts Csk. near side	
Shop Bolts Csk. far side	
Shop HSFG Bolts	
Site HSFG Bolts	

Where holes or rivets are required to be countersunk both sides they should be indicated and noted clearly.

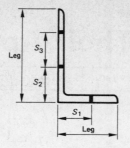

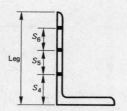

RECOMMENDED BACK MARKS FOR STANDARD ANGLES TO BS 4848:Part 4:1972

These angles are those metric sizes selected, from the full list recommended by the ISO, as British Standard Metric Angles.

Note that HSFG bolts may require adjustments to the backmarks shown due to the larger nut and washer dimensions.

Inner gauge lines are for normal conditions and may require adjustment for large diameters of fasteners or thick members.

Outer gauge lines may require consideration in relation to a specified edge distance.

Nominal leg length	Spacing of holes						Maximum diameter of bolt or rivet		
	S_1	S_2	S_3	S_4	S_5	S_6	S_1	S_2 and S_3	S_4 S_5 and S_6
mm	mm	mm	mm	mm	mm	mm	mm	mm	mm
200		75	75	55	55	55		30	20
150		55	55					20	
125		45	50					20	
120		45	50					16	
100	55						24		
90	50						24		
80	45						20		
75	45						20		
70	40						20		
65	35						20		
60	35						16		
50	28						12		
45	25								
40	23								
30	20								
25	15								

Note: That with the back marks and cross centres quoted the maximum diameter of fastener stated conforms with Table 21 of BS 449: Part 2: 1969.

This will generally result in the most economical connection.

Weld Symbols and Their Application

TYPE OF WELD	CROSS-SECTION	SYMBOL	APPLICATION OF WELD SYMBOL
Fillet 6 mm leg length	Other side / Arrow side		Other Side / Arrow Side 6
Square butt			
Single-V butt	Arrow Side / Other Side		Arrow side / Other side
Double-V butt. Top dressed flush			
Single-U butt with sealing run			
Double-U butt			
Single-bevel butt			

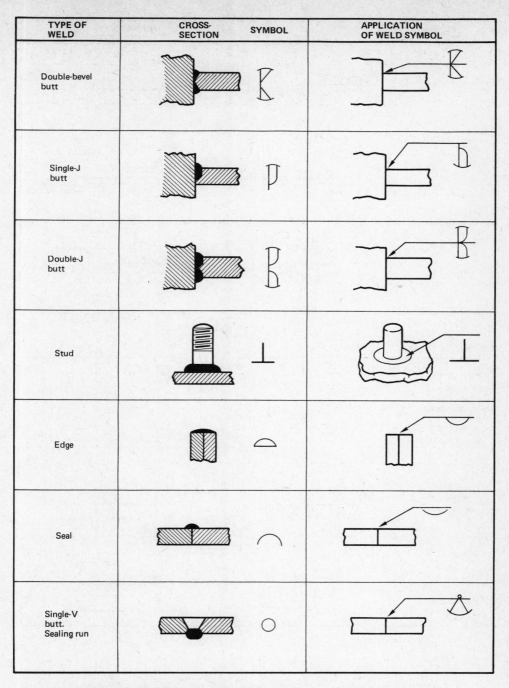

TYPE OF WELD	CROSS-SECTION	SYMBOL	APPLICATION OF WELD SYMBOL
Double-bevel butt			
Single-J butt			
Double-J butt			
Stud			
Edge			
Seal			
Single-V butt. Sealing run			

NOTE   Symbol BELOW reference line = Weld on ARROW side
Symbol ABOVE reference line = Weld on FAR side
(Cruciform sections: far side = far side of abutting plate)

TYPE OF WELD	CROSS-SECTION	SYMBOL	APPLICATION OF WELD SYMBOL
Backing strip	(a) (b)	=	(b)
Spot		✕	
Cruciform fillet		◁	

ADDITIONAL SYMBOLS

Full penetration butt weld by a welding procedure to be agreed	As stipulated		to suit
Compound welds	RADIOGRAPHIC EXAMINATION		
Fillet	WELD ALL ROUND WELD ON SITE		

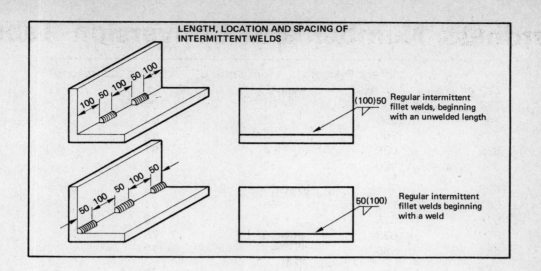

LENGTH, LOCATION AND SPACING OF INTERMITTENT WELDS

(100)50 — Regular intermittent fillet welds, beginning with an unwelded length

50(100) — Regular intermittent fillet welds beginning with a weld

Hardness Number and Conversion Table

Brinell No.	Vickers Pyramid No. (Hv)	Rockwell "C" No.	Shore Scleroscope No.
780	1150	70	106
745	1050	68	100
712	960	66	95
682	885	64	91
653	820	62	87
627	765	60	84
601	717	58	81
578	675	57	78
555	633	55	75
534	598	53	72
514	567	52	70
495	540	50	67
477	515	49	65
461	494	47	63
444	472	46	61
429	454	45	59
415	437	44	57
401	420	42	55
388	404	41	54
375	389	40	52
363	375	38	51
352	363	37	49
341	350	36	48
331	339	35	46
321	327	34	45
311	316	33	44
302	305	32	43
293	296	31	42
285	287	30	40
277	279	29	39
269	270	28	38
262	263	26	37
255	256	25	37
248	248	24	36
235	235	22	34
223	223	20	32
212	212	17	31
202	202	15	30
192	192	12	28